The Farming Story

by

Eric Chase NDA

About the Author

Eric Chase together with his brother was a tenant farmer in Hampshire for many years.

The tenancy of the farm was taken over by his father in 1925. After University Eric worked for the Berkshire War Agriculture Committee as an Assistant Livestock Officer and also involved in Combine drilling of manures.

Although young he experienced the 1930 agricultural depression and then the war years. After returning home from Berkshire in 1947, he and his brother took on the major job of the revival of the run-down 250 acre farm which involved re-draining the whole acreage. They gradually took on other tenancies from the estate increasing the size of the farm to 950 acres.

In 1952 Eric won a scholarship to visit American farms and research stations. On his return he then shared his knowledge with other farmers and over the years he developed various livestock enterprises on the family farm from dairying to beef production. He was an active member of Romsey Young Farmers Club for many years, eventually becoming President.

During the 1970s and 1980s he was involved with the NFU as Hampshire Chairman and then as delegate for Hampshire at the NFU Headquarters in London. He was also involved in the development of various syllabuses for the Agricultural Training Board.

He has always had an interest in archaeology and attended night classes at Southampton University. He retired from farming in 1987.

Published by Whiteparish Historical & Environmental Association

ISBN no. 978-0-9537744-1-8

Design layout, origination, printing and binding by
Salisbury Printing Company Limited
Greencroft Street, Salisbury, Wiltshire, SP1 1JF

Acknowledgments

The author would like to thank the following for their help and encouragement with the writing of this book.

Tony Giles – Emeritus Professor Agricultural Economics, Reading University

My wife Margaret.

My daughters - Sue, Dorothy and Hilary

Contents

CHAPTER 1. INTRODUCTION ... 3

CHAPTER 2. THE FARMING DAWN ... 6

CHAPTER 3. THE ANIMAL FARMING PARTNERSHIP ... 15

CHAPTER 4. THE BEGINNINGS OF MECHANISATION ... 23

CHAPTER 5. THE CHANGING ROLE OF ANIMALS ... 27

CHAPTER 6. THE COMING OF SCIENTIFIC AGRICULTURE ... 31

CHAPTER 7. THE CHEMICAL IMPACT ON FOOD PRODUCTION ... 39

CHAPTER 8. THE BEGINNING OF THE MODERN ERA 1850-1990 ... 42

CHAPTER 9. THE CROPPING CYCLE OF THE NEW ERA ... 58

CHAPTER 10. THE DISCOVERY OF THE NEW LARDERS ... 71

CHAPTER 11. BRITISH FARMING CHANGES COURSE ON THE ECONOMIC HIGHWAY ... 82

CHAPTER 12. WORLD WAR 1 AND FOOD ... 91

CHAPTER 13. A REFLECTION ON THE FARMING CHANGES IN THE 1930s. ... 98

CHAPTER 14. THE WAR YEARS OF 1939 TO 1945 ... 111

CHAPTER 15. THE CHANGING EMPHASIS TO LIVESTOCK ... 120

CHAPTER 17. THE TIME OF NEW IDEAS AND IMPROVEMENT OF OLD ONES ... 135

CHAPTER 18. 1957 THE TREATY OF ROME, THE WAY FORWARD FOR THE EUROPEAN FARMER ... 147

CHAPTER 19. WHAT OF THE FUTURE FOR BRITISH FARMING? ... 152

CHAPTER 1. INTRODUCTION

If you research into the history of agriculture you will find far more in the libraries of the world, than you will within this book. However, if you want to have a general view of how farming began then read on and hopefully this will give you a taste, and inspire you to read in more depth about how farming reached its position today.

These days in the year 2013, the press is full of agricultural news. Very little, in my view, is giving the background to the farming story .A lot of information has been edited into small bites, that in no way reflects what history is telling us. This often causes the reader or listener unnecessary stress. Few folk will take or have the time to read information in depth. We are all subjected every day, every hour, to sound bites of news or brief TV shots of events and these practices can and possibly do influence opinion. In my view this has become a potential danger to democracy as we know it - masking proper debate on subjects of great importance and farming is a case in point.

Farming has been pushed from pillar to post in this way, over such subjects as animal production, organic farming, use of chemicals, low wages, sprays and over production, let alone too great an influence on politics.

The way we govern ourselves today was set in motion by ordinary men and women, not brainy intellectuals. These would have been craftsmen, yeoman farmers and workers of the 17th and 18th century, who debated and worked hard for a better life.

Why, you might well ask, is this story being prefaced with a reference to democracy and free speech? Well, I think it is has a purpose with regard to farming in Britain. The average family in the UK is now many generations away from working on the land. What they do learn about farming the land often comes as I have mentioned from edited news. It is so often taken out of context and thus is potentially dangerous.

Without food and land from which to grow it, we die. It is the very basis of our existence. Our friends in Europe, because of the way they reached the industrial revolution and the devastation of wars and blockades, have an inbuilt knowledge of this fact. Those who have experienced hunger and death from starvation as many Europeans have, never lose respect for the soil.

The export of our railway knowledge, from our industrial base in the 1800s, led to the opening up of the prairies in both the USA and Canada and since then our nation has lost this feeling for the land. Our factories enabled us to buy our way out of famine and through our industrial output the Empire, as it was, fed our nation.

Things have now changed. The world is shrinking and famine is on the increase - famine in the sense that most of the world could eat better or more. We no longer have an empire "to milk". We must pay the price for food in the world market place with more and more nations bidding for produce in the same market. The population explosion, with the world population estimated to double in the next thirty years, is making us think about the land, the environment and how we are going to handle the problems that are always coming along.

The popular opinion that Agriculture is a simple operation, managed by and achieved by simple people, is an opinion that belies the truth. The subject is huge and complex and no part can be taken out of context and used to further confuse people about their food.

Farmers have always found a way of moving forward to produce food that populations want, and these days demand, certainly in the affluent west. This latest emphasis will be sure to take into account the need to work with the environment rather than against it for economic reasons. In the not so distant past, about 200 years ago, the use of scientific knowledge began to overtake inherited knowledge, of crop production. Up until then observation of what worked and what did not was all that man had.

Now we have a problem, for the pace of research in the last few years has outstripped the average man's ability to keep up. Faith in science has been badly shaken by various crises such as BSE and bird flu. So shaken, that research into genetic modification is being forestalled to such an extent that field work set to prove these theories is being held up. Western man is now so well fed that he is unable to look to the future as to how the doubled world population might be sustained. He considers that the environment must not be violated even though these new discoveries may enhance it.

By reading this book, perhaps the reader will gain a background knowledge of the past, and will be better placed to look to the future of farming and the land. This is my attempt to give that base.

CHAPTER 2. THE FARMING DAWN

Until 150 years ago, the bible was the basis from which to work out the date of the beginning of life on earth. From bible reading, and a certain amount of "bishop's licence", people came to believe that life began only some 5000 years before the present day. Mind you, this belief was already under scrutiny by amateur archaeologists, who were digging up bones of immense size from strata or rocks that appeared to be so much older. These bones were from animals that existed in the warmer climate of pre-historic times. However, at that time the Church insisted that animals and man were put on this earth by God looking as they do now. Q.E.D. No arguments.

Darwin, who had been around the world on a fact-finding mission in the 1840's started to undermine the bible theory. His book on the theory of evolution caused great debates in the western world, but has now been accepted. Perhaps not quite world wide, for some of the southern states of the U.S.A. still contest his writings.

When carbon dating was developed, it became clearer year by year that the world was of an immense age and bones of dinosaurs could be millions of years old. As far as I am aware evidence of early man was not realised until a Mr and Mrs Leakey, digging away in a remote part of Kenya, came across bones of an animal that was like an ape but not quite. Carbon dating suggested that this person was about 2 million years old, and had been walking upright.
More and more work has gone on as more skeletons come out of the earth's cupboard. The story has developed further, as it has become clearer that man's evolution, as it moved away from the apes, had many branches of which only one became Homo sapiens.

To give an idea of what is thought to have happened, I'll set out the current thoughts on the subject. However understanding may well have changed by the time this goes into print.
It would appear that some 3 million years ago some apes that at that time only lived in the trees, started to come down from them, and began to walk upright. Logic says there must have been a reason for this and climate change is considered to be that reason, possibly coupled with a genetic change enabling them capable of handling the problems that climate change had brought about.

The hypothesis is that grass began to take over from lush tree cover, and those who could cope with this new development by standing tall and walking, began to adapt to eating things other than tree fruits. Studies of these skeletons show that one chap seems to have dominated at this new juncture. He was a big, tall fellow carrying a large stomach, indicating the ability to digest large amounts of vegetation. His brain, although larger than his brother ape, was small compared with later versions. However his jaws were large and powerful.

A further 500,000 years of refinement brought about yet another change. As his brain became larger, artefacts from archaeological finds suggest that he made simple tools and used them to cut up and eat other animals. He may not at this stage have caught his meat on the hoof, but probably scavenged what he found lying about on the grasslands. The interesting thing about this model was the fact that, as his brain size had increased, his stomach size had decreased, and he was starting to migrate out of Africa. Was this change due to an ability to cook meat, thus improving the digestive process? Around this period man had begun to reach Europe and possibly to the eastern areas of the world, China and South East Asia.

Experts in this subject, reckon that when model number two got his hands on these tools, which were made of flints and could catch his meat, he had no need for a large digestive system, for his new diet was very calorie rich and parts of his alimentary canal, for example his appendix, began to become redundant. The extra calories were just what the new brain needed as for the brain to work takes 20% of our energy output.

So as more meat was eaten, and in time cooked aiding digestion, so larger grew the grey matter. The tools to cut up his food were improved to become as efficient as a steel knife today. These were the flints found in many areas which if split, or knapped, proved to be very sharp.
As time passed, these new men further developed and spread to all the Americas and Australia. Whether he was chasing animals, that due to climate change had to follow the rain, no one is too sure, but over many thousands of years new man must have begun to over reach himself, or over breed, and run out of things to catch. Chuck in a few Ice Ages for good measure and no doubt the topic of conversation around the camp fire was "What are we going to do about our next meal?"

Nothing would have happened very quickly, but with the new model's brain, observation must have told him that, rather than picking berries and wild grass seeds to go with the meat, he might well be able to grow these things. Digestion and how to improve it must always have come under scrutiny; uncooked wild wheat would not be too easy to digest. This is where the daydreamers came on the scene, and the new tool to be developed was the grinding stone. Powdering the wheat made it better for the digestion and released more calories to feed the brain. Perhaps there was a cunning plan afoot, because it appears from the observations on the skeletons of the womenfolk of this era that it was they who did the grinding. Was this because the chaps reckoned that they had to spend too much time hunting and could not help with this chore?

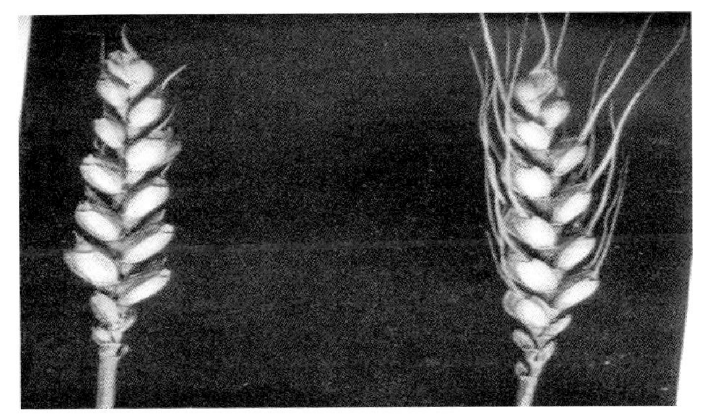

An Early Wheat

With a lack of animals to catch, meat consumption must have been on the decline. You can if you wish, imagine a group discussion going on around the campfire, and a rather brainy chap getting the elders to accept his idea of staying put and growing food.

"We have got to get hold of and keep as large an area of land as we can. Then we can burn off a section each year and grow these wild wheats in the ash. It works well in the wild after a lightening strike." He probably didn't go into too much detail such as needing an army to hold the ground, and needing to make new tools to work the soil, in case he put them off the idea. However, if this ploy worked it would also get him out of going on boring hunting trips. How or where this new development began no-one is sure, but there is a suspicion that it may well have been in Iraq.

If this brainy chap had been able to write, he might have set out his ideas in a "green paper" rather as a consultant might today. This is a cut down version of his plan:

1. Find a good level site, lush with trees and vegetation.
2. Using the new axes that have just come onto the market, cut down all growing material.
3. Leave to dry during the summer months.
4. When dry, set fire to the whole area.
5. Collect wheat grass seeds, the best and plumpest possible.
6. When spring comes, cast them into the ash, covering them using the newly invented hoe, or appointed stick.

Stick used for Cultivation

7. Harvest these new wheat plants before they drop their seed.
8. It is recommended that the elders do a rain dance to make sure of success.

This method known as the hunter-gatherer system is still practised to this day in parts of the world, such as the sub-Sahara lands in Africa, but on a small-scale. It was a great improvement on the old system of either chasing game or moving domesticated animals to new grazing grounds. But it must be remembered that moving animals to new grazing is still practiced in highland parts of the world. This is assuming that animals at this stage in man's development had been tamed or as we say, domesticated. The cropping could be repeated for several years on the same ground,. However, after three years or so, the crops would begin to fail, and the whole process had to be repeated.

There is no doubt that it worked well, but there had to be a constant search for new areas to burn. It is highly probable that as can be the case today, some folks would have been against this new development. The new axes and hoes had changed the status quo, for it was apparent to the brains of the time that things had to alter but people refused to accept change. You would now have to hold on to your territory, not continually look for new ground, because other tribes were doing the same and things could get ugly.

It is probable that this system of farming, a variation of the hunter-gatherer, went on for many thousands of years. However, as the population grew, as one assumes it must have done, this method had to change again. The vegetation that provided the fertility to grow wheat would have become more difficult to find, and the recovery time between the slashes and burns meant that there would be less to burn off and thus less residue for the wheats to feed on, i.e. less fertility available. The original thinker must have long left this earth before people began to wonder why his brilliant idea no longer worked, as well as it once did, but as we find to this day, as one thinker goes yet another comes along.

This next expert must have understood the reason for falling output, that is, lack of food for the wheat plants. Perhaps the wheat had undergone a certain amount of selection to improve the yield, and needed more fertility. Who knows?

Meanwhile, the women who had the task of pulling the pointed stick, and the new hoe through the soil found this very hard work. Gradually as the men had no longer any excuse to go hunting, because there was little to hunt, they found themselves working in the field. Not as much fun as hunting and it also interfered with the thinking process. Not for long though, for whilst leaning on the hoe, much as chaps to this day lean on their shovels, a brilliant thought came into one chap's mind. Why not train one of his recently tamed wild cows to pull the pointed stick? This idea also led to the next development which was catching fertility and putting it back on the earth.

At what time the changeover from human to animal pulling power came about is very difficult to pinpoint. No doubt the powerhouse (ox) got its fodder at night grazing around the cultivated area, and then was caught and put to work during the day. Again it is not hard to imagine that these beasts raised their tails as they worked and deposited manure and urine on the cultivated earth. The resulting increase in growth that must have occurred could not have escaped the attention of any thinking man. We can see the results today in grass fields that are grazed, where there are isolated groups of denser patches of grass. This is from the extra nutrients picked up by the animal in one place and concentrated in another.

Another husbandry pointer, that I'm sure the people of the past would have noticed, was that if ground is left for some years, you can come back and reap another harvest of fair proportions. Why this should be was not understood until the Victorians came on the scene with the science that enabled them to unravel the relationship of plants and the soil.

I have tried to convey to you, the possible way farming began to take shape in the prehistoric past. The seeds of modern agriculture were set then and have gathered pace ever since. However, at times, due to changes in climate or through man made problems, there have been long periods of inactivity, or lack of progress.
The farming "bug" may have begun to bite you, if not its history. You may be aware that archaeology has changed from being an amateur pastime, to one of high science. Now flesh is being put on the bones of the buried, and great efforts are made to try and understand the mind of our forerunners. It is not an easy subject but one in which science is opening windows through which we can look.

Wheat, that cereal that is still our staple diet, in the west if not the eastern world, is bound very tightly with early man. The 'Brain' that started to pick and eat the wild wheats and moved them into his own fields, by selection, would be amazed at what he started. Those first handfuls have grown over 7-8,000 years to yield millions of tonnes each year. That's a lot of dough one way or another! It looks to me that if 'Brain' and his wife had not started up the DIY method we now called farming, we would still be fighting it out on the grasslands of the world rather like the lions, with teeth to match.

In addition to this is the point that if he had not used the animal power he also would not have made further progress. Without animal labour a family can only cultivate some 1- 2 hectares at the most for their own consumption. No extras for sale, or time to cultivate other interests, the use of animal labour would have doubled this area or alternatively, given them the opportunity to think of other things!

CHAPTER 3. THE ANIMAL FARMING PARTNERSHIP

As we saw in Chapter 2, chasing something edible became a very uncertain way of staving off hunger. It worked all the time the animal population outnumbered the human. No doubt climate change had a great influence on the animal population, this taking place over very many thousands of years. Maybe this was one reason that early man moved from Africa into Europe and beyond - by following the herds. As the trees failed man had to learn to walk. Having no tree fruits to live on they developed a taste for meat, simply because grass was the only thing on offer. One day we may well know what took place, but for now we can only guess.

The problem that must have happened in this chasing game, was that groups of folk must have been continually bumping into others, all after meat. At this stage, can we not imagine yet another Brain coming on the scene, who could see that four legs must be better than two when it came to hunting? Speed was the way forward, and the horse was the animal to provide it. He, this clever chap, must have found the way to get command of the horse. The only evidence we have today of how this must have happened, comes from the Red Indians of North America, who took to the horse, to be able to follow the buffalo. Although in fact they actually pinched the feral horses roaming the prairies from the Spanish invaders of Mexico, and then taught themselves to ride, without saddles!

All those who watch western movies, will have a fair idea of how horses are broken to obey the commands of the rider. It is a case of tiring the animal until his will to resist gives out. Now the cunning part in all this is not to break his spirit, and this is where the psychology of horse breaking comes into play. Man found that with quiet movements of hands and legs he and the horse could become as one and suddenly his speed increased from ten mph to thirty, and could be kept up for long periods. At the time it did not further the cause of agriculture, but out manoeuvred the chaps who failed to take advantage of the new method. No doubt it was good fun!

The Greeks in the Bronze Age period produced a manual on horse breaking and management that cannot be bettered to this day. The ideas that were applied to the horse were also applied to other animals, bovines and dogs, and we must not forget the camel. All of these except the pig could be ridden, milked and eaten.

There is a rather nice story coming from Chinese folklore, I believe, that suggests the way the pig became a meat provider. The story goes that one day wild pigs became trapped while foraging in an abandoned hut. The hut caught fire, roasting the animals. A family passing smelt something good coming from the smouldering hut ,and took a taste of the roast pork, and were 'hooked'.

Another idea on how the pig was domesticated also springs to mind. As the wives began work on growing things, they no doubt found it hard to keep wild pigs out of the crops. They too came up with a cunning plan. They may have told their men to catch them some wild pigs which they would tame. Then they would produce young which would be good to eat. Maybe, they then suggested to their men folk that, as they found hunting so time consuming with little reward, they could now stay at home, at least during critical periods of the season, i.e. sowing and harvest time.

The turn of the chicken also came along at about this time, and the hunter-gatherer period started to come to an end. It goes without saying that not all of the world's population had learned of these new ideas. It depended on where in the world you found yourself and what climate predominated in that area.

To this day, there are still pockets of people who still hunt and gather food, for instance , the indigenous folk of Namibia. They have adapted to live in one of the driest parts of the world, amongst animals that have had to learn, no doubt by natural selection, how to manage drought. However even here, in the last hundred years, outsiders have muscled in and started to farm sheep on these arid acres, and it does not look as if the true locals are going to be able to compete. But, they have set modern man thinking about how the future of food production will develop, in the more challenging parts of the world.

The Eskimo also still clings on to his hunting way of life but has also started to rely on alien foods that have come his way from modern farming methods. And who can argue with wanting an easier life? Buying food from a shop is less unpredictable, and simpler than waiting in sub zero temperatures for a seal to pop his head out of a breathing hole.
Even the fish of the rivers and seas are being farmed, because the wild fish stocks are so low, through over fishing or hunting which is the reason that fish is so expensive these days. So what you see, before your very eyes, is only an indication of what has been taking place over thousands of years.

Growing plenty of food has meant that the world's population has been able to increase, which could not happen under a hunter-gathering system. Man has also been able to travel up other roads, the arts, invention and exploration that abundant food allows them to do.

The stages in this farming story have moved from hunting, then slash and burn, alongside animal co-operation, to the start of an era of understanding the soil, which appears to be around the time of the Bronze Age. The work by historians on the Bronze Age is gathering pace, and the fortunate discovery of a frozen body in 1992, high up on the Austrian Italian Alps, has helped us understand these people. Not only did they have sound knowledge of weapons and warm clothing, but they also had settled farming systems down in the sheltered alpine valleys.

It would appear that they managed their livestock by using grazing high up in the mountains during the summer, and this suggests that they could winter their stock by making and feeding hay. If animals were kept in this way, they must have produced manure which would have been put back on the lowland hay fields. It seems very probable that the farming systems still in place in these mountain areas of the Alps stem from this Bronze Age time.

Apart from this, the Bronze Age man's knowledge of woodland management, together with his use of animals to keep the ground cleared, has been well researched. No new tree shoot could sprout before a sheep, or cow, would nibble it off and thus kill the seedlings. This system can be overdone however, and, under dry climatic conditions, can cause deserts to spread. The goats of Africa are past masters at this task, but if they are kept under control, and new trees planted to help speed up the return to good soil, the deserts could bloom again.

In England, it is also possible to show how Bronze Age man cut, or pollarded the trees so that they would grow fresh shoots. These shoots, after about 30 years, would have produced timber that a bronze axe could manage. As Bronze gave way to the new material, Iron, trees were felled left, right and centre as more ground was needed to grow cereals for the ever growing population.

At this time it looks to me that farming was changing and the germ of economics was beginning to appear. We have only Roman history to guide us in this, for they

were the dominant Iron Age tribe in Europe bar none, having mastered the science of economics, trading and politics.
A few hundred years, after they had established the City of Rome, and subdued all of Italy, they found that they could not grow enough grain in their country to feed the urban masses. Unrest was the last thing a Roman Emperor wanted, so grain was imported from their conquests in North Africa. Free parcels of grain were given to the population. In time, as the Empire grew, and the Rhine frontier became established, a large army was needed there to keep out the war like tribes from across the river. This army needed feeding and as a result Britain began to export cereals to them. In exchange for cereals and hunting dogs, came fine pottery, wine, and more iron tools.

England is covered with Iron Age hill forts, as are France and Germany. The idea of these forts was one of defence. Should your area be attacked by a rival tribe your kith and kin, together with your animals, would come into the well fortified area and could eat drink and be merry until the enemy got fed up and left. The hill forts became bigger and more sophisticated as time went by, until the Romans came along and defeated these tribes, took control of the farming and pushed them into a new era.

We know from the work that has been done at Butser Hill in Hampshire, that these early farms could produce nearly a half a tonne of wheat per hectare[1]. They used oxen to pull their pointed plough sticks, and grew a variety of wheat that grew tall, and thus away from the weeds. Sheep were the dominant animal, producing wool and meat, and no doubt milk as well. The breeding animals could survive the

[1] 1 hectare = 2.471 acres.

winters grazing what little grass growth there was. When spring brought a flush of growth, the ewes would lamb, and the grass would be turned into flesh and milk.

These men had begun to master a method of keeping the fields fertile. A patch of ground would be cleared, and on it they would grow wheat, the next year barley, and the third year leave the field fallow. Then another field would have been put into rotation and the same method of cropping applied to it. The fallow field growing weeds and grasses would be grazed by the animals. Thus minerals came from above and below. The wheat crop would be sown after the year of fallow, to take full advantage of the renewed fertility, for wheat is a greedy plant.

With an excess of food, other things than farming could be undertaken. The Romans, who borrowed heavily from the Greeks, built roads and great buildings, tamed the rivers, and improved health. They brought these benefits to Britain, together with a few new crops to help with the old rotational system of the local tribes i.e. legumes – peas, beans and clovers which could trap the essential nitrogen in their roots that the crops craved.

Having dealt with, the origins of farming, I will now refer back to the Roman influence on British Agriculture. Although they failed to produce enough food for themselves, they were great ones for putting into operation other people's ideas. For instance, they followed the Greek civilisation in many ways, and copied farming techniques from their far flung Empire.

After the final overthrow of British resistance, it was the retired Roman soldiers who probably gave the next lead. It was the practice of Rome to retire these veterans after 25 years' service and to give them land to work. The expansion of the Empire was in part to get more land to put these chaps upon. There were always a lot of men finishing their stint, a promise was a promise, and to keep a fighting force keen, promises must be kept. The first veterans' small-holding system was in and around Colchester, where there was good land, a big army camp on the spot along with a river, and close to the sea.

All went well until they mishandled Boadicea, a country girl at heart, but that's another story. But that aside, these men, who had been fighting all over the Empire, had brought some seeds along in their pockets, picked up a few ideas and now had a chance to try them out. So in came things like carrots, parsnips, cabbages and the vital one, beans. They were good to eat in the winter, when dried they kept well, and as if by magic the crop that followed was always good. This legume, along with others such as peas and clover, fixed nitrogen from the air, and was the next breakthrough. Legumes were used to the full until the 1950s, then a lapse occurred until the 1990s, but it may well be that now the legume is to make a come back.

Varieties of Clover

The reason legumes, such as clovers, began to be used less in modern farming rotations was due to oil. Nitrogen could be made from oil very cheaply and it became possible to add extra to the soil, at a rate higher than legumes could provide.

Various varieties of crops such as wheat were developed to use this extra food and thus yields increased. It was known as the 'green revolution'. It may not be sustainable but we shall see!

Instead of a pointed stick used for cultivation, the Roman plough – a great invention, was also a leap forward towards better crops as we shall see in the next chapter.

CHAPTER 4. THE BEGINNINGS OF MECHANISATION

I have recounted how I thought someone had probably used a bullock to operate the pointed stick plough. This particular cultivation was almost certainly a dusty experience, and as the stick moved so little soil, the operation had to be repeated many times. Very old field systems have been uncovered by archaeologists showing the many times the ground had been moved to get a seedbed or tilth. If this was not done the resulting crops would have been even more disappointing than normal.

This problem would probably have been discussed many times around the fire. What they wanted to achieve was a fresh surface of soil, a new beginning, similar to the virgin soil after it had been cleared by burning. If this could be done, then it would also go a long way to pushing the trash out of sight, and thus keeping down the weeds in the next crop. It also should save time for other work, or possibly just dreaming and drinking!!

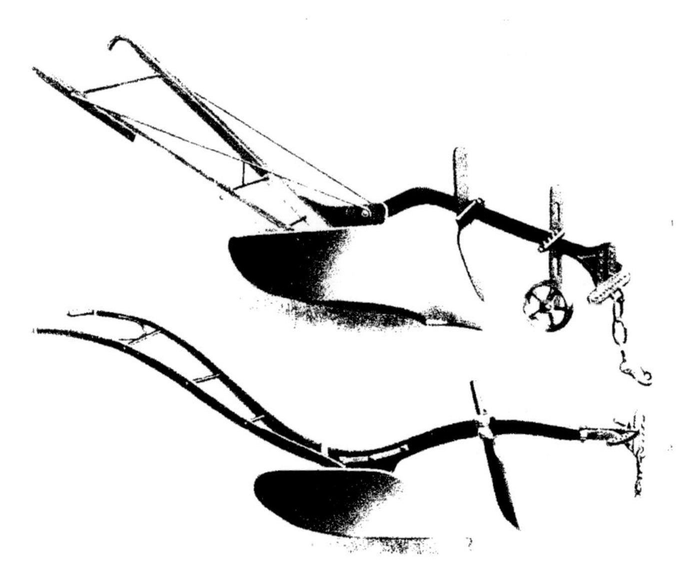

An illustration of Mould-boards for turning soil over

When this next breakthrough came I am not sure, but it must have come in around the middle Iron Age period, for the Romans seemed to be using this new idea. The technique of metal working had progressed at a great rate, more iron ore was being found and smelting was proving easier. Thus along came the plough, albeit still based on that pointed stick. The point of the stick was covered in metal and a sloping platform also made of iron ran back from the point or "share". As the share cut into the soil the earth moved up the slope, and was turned upside down thus presenting a new surface. The power needed to move this plough along was much greater than that used for the pointed stick, and needed at least two cows or donkeys. It seems the horse at this stage may not have been used. It also could reasonably be assumed that the extra animals now involved would have increased the output of manure, thus helping fertility.

The above demonstrates the yoking of oxen to increase the pulling power

An interesting point has come to light from work at Butser Hill farm in Hampshire. An attempt has been made to demonstrate how Iron Age farming may have worked. There researchers discovered that when they planted some of the oldest wheat varieties they found that, using the stick system, the wheat plants which grew to some five feet tall, yielded well. In fact they recorded yields of two tonnes per acre, which is only one tonne short of the 1990 average of 3 tonnes in the UK. This was because these plants grew away from the weeds, so all was not bad with the old system. Even around the medieval period they could not reach these yields. Why this was, is difficult to find out. Maybe the press of population was taking too much out of the soil.

To harvest the cereals a sickle was used for many thousands of years. This was a back breaking job, but eventually another clever chap invented the scythe, which was a two handed tool, that allowed you to stand upright and cut an acre a day. The slow, easy, smooth motion of the scythe was to become the harvesting method for many centuries. Then the exodus of land workers during the industrial revolution, and the increased demand for food, brought along the horse reaper.

CHAPTER 5. THE CHANGING ROLE OF ANIMALS

Who was the man I wonder, who in the dim and distant past realised the useful tool of castration? It must have taken a considerable amount of medical knowledge to realise the connection between testicles and behaviour. Did this realisation begin with animals or man? It certainly was the case in the changing role of animals within agriculture. As far as one can tell no progress was made in the improvement of animals by selection although this assumption may well be proved wrong by archaeologists in the future, using the new scientific methods such as DNA testing.

Common land, which must have been used for grazing animals of both sexes, meant that the animals could breed freely. So there was no opportunity to do any selective breeding to improve the stock. The selection that did take place must have been at the behest of the climate and foodstuff available i.e. natural selection, the survival of the fittest, or those who could get through the winter months.

Conservation of crops for winter certainly took place during the Iron Age, at any rate according to Roman history. The ability by the Roman military to keep cavalry horses going during the winter proves this. The animals would have given birth in the spring, and the young suckled on their mothers all summer. Their milk dried up in the winter, because of the lack of sufficient food to generate the supply.

When the calf was about four years old, its mother was probably killed. The new young beast would have been kept for work, and the old one salted down for the winter food store.

Of course this system also meant one less animal mouth to feed during the winter and a little more variety of food for those people in the hut. This method lasted until the Middle Ages. Then, if they wanted fresh meat, and I bet they did after all that salt beef, they may have gone deer hunting, because deer live in woods where cattle fail. Woodlands were for many years under the control of the King and his friends and relations. The history of the New Forest illustrates this point very well. This is worth reading about.

The Norman conquerors brought the rabbit, and the keeping of fish ponds. What the average man had to eat in the way of protein goodness knows. His pigs on the common and a bit of poaching on the quiet possibly helped. Until the press of population came along, this seemed to keep most people happy.

Of course the Black Death, in 1350 and its reappearance during the next 100 years, was a nasty jolt to the arrangement. However in not so many years Homo-Sapiens increased alarmingly yet again. The method of food production had to solve this challenge, and so it did.

In passing, it is I think, interesting to look overseas, to understand what was happening in the way of breeding improvements. It was the Arabs this time that made the next advance. No doubt they were helped by the desert areas they lived in, and the political system that existed. The local ruler held domain over oases surrounded by desert land. These green areas, that had barriers of desert land for defence, contained his herds of goats, camels and horses. Horses were the war machine that kept his tribe safe, the camels were used as beasts of burden. If you could breed better horses, to your specification, then you were in control.

Artificial insemination was his brainwave - something that was not exploited in England until the 1940s. A stallion of proven ability was chosen and semen was taken from him and used on the best mares. By this method he could cover a greater number of females than by natural service. All inferior males were castrated. Being a law unto himself the Sheik could really make progress and what an improvement that proved to be! The Arab horse has great stamina, can achieve long periods of work with little food or water. He became the basis of our racing stock, but that is another story.

An Arabian stallion

To a degree Britain was isolated, being an island. It also had valleys, and highlands in which animals were in comparative isolation, so specific breeds began to appear, especially sheep, but they failed to improve beyond the climatic restrictions in the area.

This was not good enough for the new demands being put on farming. It was not until the Enclosures Acts came in i.e. from the 1500s to a final fling during the

1800s, that things began to move forward. Selection for early maturity for market, and weight increase etc. was then possible. This was all because the herds and flocks could be kept under control.

Between 1710-1795 cattle sent to Smithfield Market, London doubled in weight, due to breeding, though we must not forget that better fodder crops also played a part in this story.

CHAPTER 6. THE COMING OF SCIENTIFIC AGRICULTURE

Unwittingly science, or the understanding of the fundamentals of life, was having its debut when man first began to realise that something in the soil was needed to make the plants thrive again. As growers decided to return plant and animal by-products to the soil, these contained that certain something, although they knew not what.

The first natural chemical to be used was chalk or calcium carbonate $CaCo_3$. Someone in the mists of time found that crops responded well if the chalk was spread on the soil. As it is widely found over most of Britain, they dug it out creating holes or chalk pits over the countryside.

Besides helping the cereals it had an effect on the working of the soil. Heavy wet clay land became easier to work, giving the small seeds a cosy bed in which to start life. What was not realised, before the 17th century, was why this chemical should be so beneficial. It was the lot of the early chemists to discover the leaching power of rain water, and what it took out of the soil.

This leaching of the soil makes it acid and the calcium carbonate counteracts this. Of course in other parts of the world, plants have evolved that can grow in an acid soil. In the British Isles, in the high rain fall areas of the west, grasses have become established which tolerate the acidity. The snag is that they do not give the yields that grasses on alkali soils do.

It was the demands of a hungry population that triggered off the need for more food to be grown. In the 12 and 13th centuries, the only way was to bring more land into cultivation. More trees were felled and the land farmed. The early

European wars of the 17th century began to push up the price of wheat as it could not be imported. The price of bread rose, especially if Britain had bad weather at harvest time. A population explosion of sorts was beginning in this land, and nearly every acre that could be was planted to cereals. In fact around the 1800's there was nearly as much land under cultivation as there was in World War 2. Now, if you cannot find anymore land to plough, the next step is to increase the yields. Try as they might this was proving very difficult. Chalk helped, beans helped, manure from animals helped, but no significant increase took place.

From this scenario things began to happen. This was brought about by the landowners of the time, who also had the politics in their court. They could see the way to make money was from the rents of their tenants. They started clubs where agriculturalists could meet and in modern parlance bounce ideas off each other. King George III, or Farmer George, as he became known, was one such man who it was said, was more interested in farming than losing the American colonies.

The opening up of South America, and the British influence in Chile, was the next breakthrough to increased production. Sea captains having taken gold prospectors around Cape Horn to California, no doubt wanted a return load. This took the form of sea bird droppings, or Guano as the locals called it, obtained from the islands off Chile. These edge the eastern Pacific Ocean where masses of fish-eating birds live. They roost on these islands and after a day gorging on fish have a great deal of waste to get rid of at night.

A climatic condition caused by the high mountain ranges close by, means that rainfall is very low indeed, thus the bird droppings stay put and dry. These deposits that had accumulated over millions of years were huge. It seems the locals had latched on to the fact that a boat load of this stuff did wonders for the crops it was applied to. Good commercial news spread fast, even in the day of sail. Chile being on the main shipping route to Europe, resulted in loads of this Guano, arriving in England.

Another find at this time, in the eastern counties of England, was Copra. Oddly enough this was the droppings of animals living millions of years ago. A farmer wondered why only parts of his farm grew good crops, and on investigation found this fossilized material. For a few years this was mined and sold for good money.

Both of these fertilizers were soon snapped up especially as both could be carried well away from the docks, by the canals and, after 1850, by the new railways.
The chemical composition of these organic manures showed a good deal of ammonia, NH_3 and Phosphate, P_2O_5.

Well, the birds had done nothing more than the Brain's animals had, coming home at night and bringing fertility from one place to another. In the Middle Ages the emptying of town 'privies' or toilets on the big open fields was doing the exact same thing. However, no one knew why, or rather what, was causing this to happen to the plants.

It is interesting I think to jump ahead a little to see what happened when other economic conditions occurred. When the big prairies of Canada and the USA were opened up, the British farmer could not compete on the price of cereals. To overcome this, he began to feed the cheap grain to animals, and the meat to the people. So fertility was being taken yet again from one place to another, and the effects would show in future years.

The Industrial Revolution was the key to understanding the use of chemicals in industry and farming. The Germans were in the forefront, and used their knowledge to help grow more food during WW1. Coal, which drove the revolution, was the base from which many chemicals were derived. In the heating of coal for gas production ammonia is produced. Put this on the land when you cannot get any Guano, and the crops flourish for a while.

The general outcome of these new moves, in the use of imported plant food, was the coming of the hedge. Botanists can, by looking at the make up of the species growing in a hedge, determine its age and by this means they have found that many go back to Anglo-Saxon times.
Hedges were used initially to denote a boundary of a kingdom. Whether they would have also kept animals in is open to debate. Someone found the very plant you needed for that job was the hawthorn. It may well be that hawthorn was discovered during the Crusades, used in the Middle East as a corral to keep sheep safely in at night. Someone then found a way of transplanting hawthorn and controlling it by regular trimming, and when fully developed it could stop a cow from escaping. As long as a hawthorn hedge is cut each year it will last for ever. If it does get away and become outgrown, it can be renovated by layering it.

New Hawthorn Hedge

All this entails work by hand it cannot be mechanised, which does not go down well in this day and age.

The demand for more food brought the Enclosure Acts into being and hedges were the key to the operation. Oddly enough the Enclosures Acts brought more out-cry from the population than the destruction of the hedges do today. The reason was because it meant the loss of common land which was land considered to be unsuitable for cultivation. This, together with the loss of strip farming, small plots, spread all over the parish as in Norman times, denied the average chap (the commoner) his independence because he was then unable to keep the odd cow or pig on the common land. This land was selected by owners of farms surrounding it for enclosure and they had to apply to Parliament for permission under the acts, to take it over. Commoners would thus be excluded, they then lost the chance to run a cow or pig or even geese for their own benefit. The new owner would have all this to himself. For more information on the Enclosures Acts see Appendix 1.

To be fair, each area to be enclosed was subject to scrutiny by a 'commission' who did their best to be fair. They made sure small areas of land would go to the commoners. Areas of land they could call their own. Great efforts were made certainly in the later Acts to distribute small parcels of land to the landless, in compensation for the loss of these common rights.
Money however was the temptation that defeated the peasants, for the big man had only to dangle a bag of cash in front of the ex-commoners. For many this was too big a temptation and they sold their small piece of land. Thus they then depended on finding work on the new farms or starve. Independence, even if it was only owning a cow or goose, together with casual work was lost.

At last in broadest outline the way was now clear for the advancement of agriculture. Crops could be grown without the restrictions of the strip method and now livestock could be controlled. Fertiliser could be applied and one man could reap good rewards. Animals could be selected for breeding without the common herd diluting the genes, and improved cattle, and sheep breeds began to come off the drawing board.

Henry VIII it was who decreed that no stallion under a certain height could be used for breeding. This was nothing to do with increased food output, but with defence in order to increase the size of war horses. How successful this was we have no idea, except to note that the barrier of the Welsh Mountains enabled those who lived there to keep their small Welsh ponies small. No doubt Henry's activity did increase horse size in a general way, and people began to use them more and more for farm work. However it was once again the demand for food that pushed the horse into the forefront of the fight.

As more acres were needed, the speed of cultivation had to be increased. Up to this stage oxen had been used to do the ploughing. The Doomsday book set in motion by William the Conqueror, recorded all that existed in the country and illustrates this early use of oxen. It required quite a string of them to pull a plough. They had to have a lot of room to turn at the end of the field and two men to work them. Besides these problems, they had to have regular rests to allow them to chew the 'cud', and their cloven feet had to be shod with leather shoes not metal as with a horse and that was not easy or cheap.

Now the larger horses, made larger still by importing bigger stock from the Low Countries, could pull a plough at 3 mph, twice the oxen rate, and needed only half the number to do so.

A Shire Stallion – capable of pulling a tonne.

Lack of finance and expertise in constructing roads meant that they were in a very poor state. Horses could only be used as pack animals as the roads were too unstable for wagons. This was not a very efficient way to transport goods from one place to another.

A change came about with the new turnpike roads constructed in the late 17th and early 18th centuries. A single horse could then pull a cart carrying 10-20 cwt. But what was even better were the canals that where dug and gradually laced the country together. Then the horse was able to pull a boat with up to 25 tons on board. Under the influence of these new developments industry was really able to motor.

However, with the increase in manufacture came the demand for more labour, so folk went from country to the new towns. Many having lost their 'common' independence found this new development a godsend. As lifestyle improved so up went the birth rate once again - more mouths to feed at that stage but no extra food from other parts of the world at that time.

Up to this point the basic working of agriculture was in place, but it was not able to cope with the fast changes taking place without the help of the chemists and the engineers quickly arriving on the scene. So we must move on to the next development.

CHAPTER 7. THE CHEMICAL IMPACT ON FOOD PRODUCTION

As far as fertility is concerned i.e. the food for plants contained in the soil, as I have mentioned before, was not understood until the early 1800s. The essential elements are nitrogen, phosphate and potash. All are needed in varying amounts, differing in proportions for different plants.

Up to the late 1800s, apart from the manure given by the animals and humans, all needs came in the main, from such things as Guano, soot, shoddy (a waste from the woollen industry) ground up bones, and if you were near the coast, seaweed. The quantities used increased year by year and eventually, the Guano began to run out and the others were superseded by the so called artificial manures coming on to the market. More on this subject later, but the use of these new products was to do with the economics of the market place. To keep wages low was to mean more profit for the industrialists. The cry of the people was for cheaper food, and the gradual movement of workers from the land to the factories.

The theme through this story has been the ability of man to come to his own rescue when the chips are down. When a mysterious disease affected the vines of France in about 1870, ideas were very soon put into motion to solve the problem, which was powdery mildew a disease attacking and ruining the grapes over huge areas of France. In this case the great brain Louis Pasteur from Paris, managed to overcome the fungus by the use of chemicals. A combination in fact of two elements, copper and sulphur, natural but man-made so to speak. It did the job and saved France from an economic catastrophe.

A disaster had befallen Ireland in 1840. It was the potato disease blight, which led to the potato famine that changed the history of the country and helped make the USA. This blight was only overcome by using Pasteur's concoction.

The background to this Irish famine is well worth looking into, because it could show us what problems we may be storing up for ourselves in the future. It hinges on the dependence of a growing population, on one crop. In the case of the Irish it was the potato, brought in by Sir Walter Raleigh from South America in the late Elizabethan period. What a great find it was, to be sure!

With the potato, one could get a great deal of food from a small area of ground. No grinding was needed, just peel and cook, with perhaps a little salt to taste. The one climatic condition it demanded was not to be exposed to frosts in the spring, after it had thrown up the first shoots. As far as I know Sir Walter could have played a big part in encouraging the crop in Ireland. He was given a large estate in that country by a grateful Queen and from all accounts managed it well.

The climate in south and west Ireland is ideal for the humble spud, no early frosts and plenty of rain so it did well. No doubt over the 150 years since its introduction the plant had been changed by selection for greater yields.

Much of the southwest of Ireland is not the best of land, being very stony with peat bogs, but it was land on which the more humble farmer could make a living, especially with the humble spud to help. An acre of the crop could produce enough food for a large family for a year, feed a pig, and with a cow living on the peat bog lands he could live comfortably, if not well. What is not realised is that the potato enabled the population of Ireland to explode over a very few years. As it was not possible to cultivate other crops on such poor land, the collapse of the spud, was a disaster.

Each year more and more potatoes were grown, until one summer in 1840 things went wrong, The leaves started to die, and those potatoes that were under the ground would not keep, but rotted into a nasty mess. Suddenly there was no food for the winter - that dreaded fungus we now call potato blight had struck. At the time, no one knew that a fungus was responsible, and so did not know what to do to counter the problem. They had to learn from Pasteur to solve the problem, using his new ideas with chemicals. Bordeaux mixture as it was called, was the only cure, or preventative treatment for the potato blight, right up until the 1960s.

It was also not until this time, that anyone worked out what climatic conditions were required to trigger the ever present fungus into action. Eventually it was noted that a period of high humidity and temperature, was the signal for the fungus to get cracking. This was called a Beaumont Period. These conditions are now monitored by the Met. Office, which keeps farmers informed, thus saving the waste of spraying until the time is right.

This idea or development has proved most useful, for now methods have been developed for other diseases. So that blanket spraying, which is not only harmful to the environment but uneconomic for the farmer, is not required. Spray only when you can see the white of their eyes, so to speak.

CHAPTER 8. THE BEGINNING OF THE MODERN ERA 1850-1990

Although the heading of this chapter puts the modern era as starting in 1850 this is perhaps incorrect. Rather like the beginning of the industrial surge it is impossible to put a precise date on the starting line. However, it was the start of a time, when the ideas worked on some 60 to 70 years before, were put into general practice.

During the latter part of the 18th and early part of the 19th century, men were at work trying to put more science into Agriculture. I will mention a few names at this stage, because until now I have only really been able to indicate progress by referring to a 'Brain'. Now we can put names to the brainy people who began the ground work which was taken and exploited from 1850 onwards.

A chap called Jethro Tull as early as 1733, had the notion of growing crops in rows, rather than broadcasting the seed. The basis of his idea being, that sowing in rows gave the chance to hoe between the rows, thus keeping down those dreaded weeds which sucked so much out of the soil and made life difficult at harvest time.

In the 1780-90s another man came to the fore, not as a farmer, for he failed miserably in that direction. His name was Arthur Young. At this time, the government was setting up a department, which was to be the forerunner of the Board of Agriculture. I think a reason for this, was the French Revolution across the channel which was making the government of the day think about what might happen in England. A peasant with a full stomach is far less likely to think of revolution. The war with Napoleon could cut off grain supplies, putting the status quo under threat, so a way of ensuring supplies by farming improvements had to be investigated.

Mr Young achieved this, after being made secretary of the new department. He, and others, travelled to all parts of England recording the yields of oats, barley and wheat, together with the methods used to gain those yields. They also looked at animal husbandry, to give a full picture of how things were across the broad shires. Generally things did not look promising, especially the yields. So to try to improve things, they became an umbrella organisation which set out to encourage scientific practices in agriculture.

To this end, they brought in an up and coming chap called Humphrey Davy. He was really the brain behind the new era. He had started life as a doctor. But later turned from medicine to concentrate on work involved in the chemistry of soils and how it affected farmers of the day. He did work on rotations, of which more later, and it was he who discovered that nitrogen was the mystery ingredient that those legumes were putting into the ground.
From 1800 until 1813, he gave a lecture each year on the Elements of Scientific Agriculture, at the London Royal Society, the leading centre for all new scientific ideas.

Up to this time, farmers were putting back into the ground, via manures of both animals and man, more or less what was being taken out. However, from now on as the town population increased, the human manure was not so easy to return, and the soils composition was becoming unbalanced.

The high price of wheat from 1799 until about 1820 meant farmers were hunting for ways to push up yields and thus their income. War was a wonderful thing for this, and some 200,000 acres were brought into cultivation, a great deal of it enclosed from common land. Humphrey Davy was also the inventor of the miner's safety lamp. He was made a knight. After all that work he certainly earned it!

It was Farmer George, the King, who led the way in showing how the new scientific agriculture might work. He was more interested in farming than losing the American colonies. His interests in the countryside as I mentioned earlier took prccedence over political matters. His interest in the Merino sheep which he imported from Spain, gave rise to their eventual arrival in Australia, which put the then colony on its feet economically. Wool production is still today an important factor in Australian life.

A Merino Ram

After his death and the eventual enthronement of Queen Victoria, her husband, the Prince Consort took up the challenge. He was the architect of the Great Exhibition of 1852 held in Hyde Park, London. This man could see, or put into focus, how the great leap forward was going to take place, using the new tools which were coming to hand.

We are in a very similar situation today, but with out any leadership to help us find the best way ahead. We have the new tools of computing, genetic manipulation, disease control without the use of, what many consider to be, the dreaded chemicals.
In the western world we have reaped the benefits of over-production, thanks to the new agriculture being developed in 1852 but the third world still needs help to find its way. It is going to be our task to transplant our ideas to the third world's farming. Let them also march into the future as we did back then, giving all a full stomach.
Politicians will have to give us the vision and the money to do it. Will they?

What were the challenges showing up on the 1850s horizon?

1. A population explosion of gigantic proportions, look at the figures About 20 million .rising to 35 million in about 30 years;
2. The area of fertile land available but not geared up to feed this explosion;
3. The need for speedier, more efficient land work because the labour was going to be sucked off to the new factories;
4. Draught animals eating a huge amount of fodder to provide the energy to work, thus reducing the acreage of land to feed the human population;
5. The ability to recycle fertility and the need to import extra plant food from abroad;
6. The need for experimental farms to help map the way forward;
7. The need for national control of the agricultural potential.

We have looked at problem No. 1, now we shall see how problem No. 2 was tackled.
Those Romans knew perfectly well that waterlogged soil would not grow anything of any consequence. Realising their potential, they took the Fenlands of Eastern England, and controlled the rivers and drained the land, by digging huge canals. This work was continued by the Dutch through the 17th and 18th centuries. They happened to be the very best experts at the job, having cut their teeth on their own home patch. Being near neighbours of the East Anglians their expertise was soon common knowledge. However those methods for draining the Fen could not be employed on other soils.

In some of the clay soils, classed as 'heavy' by farmers and anyone else who has had the misfortune to dig them, water does not move very quickly through the soil profile to the subsoil. These heavy soils take time to dry out in the spring, and so are slow to warm up reducing the ability to get the planted seeds away to a good start.
To overcome this folk had tried another method of drainage. This involved digging a trench some two feet deep and filling it with stones or a mixture of stone and heather. This made channels for the water to flow away to an open ditch. However, after a few years the finer soil particles tended to fill the channels and impede the flow.

The next development was a cross between a clay pipe and a horse shoe. I imagine at that time it was difficult to make a fully rounded pipe of some 2-3 inches in diameter, so a horse shoe shaped clay pipe was produced and this was laid on flat tile. These pipes were of course porous, water flowed through them easily. They must have been laid 150 years ago but can still be dug up to this day. Then along came the steam and the machine age and in came the fully rounded tile drain, millions of which were produced and very cheaply.

It is only now that they are beginning to be superseded, this time by plastic.

Of course you could not lay these pipes anyhow, levels had to be taken of the field to be drained and a scheme worked out to take advantage of the shape and size of the field. Since the introduction of clay pipes land drainage has become a science. Big open ditches were dug all round the outside of the field to take the water away to the rivers. A massive amount of work was involved. Think of the huge tonnages of soil that had to be moved by hand, to create trenches 2-3 feet deep, 22 yards apart, some even closer in the heaviest clays.

To bring more land into arable production, and it paid to do so in the 1850s-60s, a land owner could get a grant from the government to carry out this work. It was in the national interest to do so, and the big estate owners still had a big political say. As it happened the labour to do the job was on hand and skilled at moving soil. They were the Navvies, so named from when they had been recruited to dig the navigation canals, at the start of the industrial revolution. They then moved on to the construction of the railroads, and as they began to run out of this work along came land drainage. Many millions of acres were drained over a period of some 30 years. The outlets into ditches were constructed of brick and often had a cast iron plaque giving the date of the work. They were so well made, that it is still possible to find them today.

Points No. 3 and 4 relating to faster land work together with the huge quantity of food consumed by the draught animals, were addressed by the coming of steam power and its application to work on the land. A slow process indeed, for it was some 100 years before the horse was finally ousted by the combustion engine. The Victorians soon had the stationary steam engine working on the land. An engine was put at each end of a large field, and then connected to each other with wire hawsers. In the middle of this was an eight furrow plough and this outfit, worked by four men, could plough some 20 acres a day.

At that time a two horse team and one carter could manage just one acre per day and no more, as horses had to rest overnight just to achieve that. So when you compare that with the output from the stationary steam engines system, it can be seen what potential was being let loose on the land.

Steam Engine used for Ploughing

Ploughing – note Steam Engine at top of picture.

Initially, most of the early steam power was put to work in the naturally big fields of East Anglia whilst the smaller fields to the west had to keep the horse. But when the combustion engine came in the hedges were removed to make better use of the new power. That caused an outcry almost as loud as that during the enclosures era, but for different reasons.

To plough that 20 acres with horses, would require 40 animals and 20 men. Each horse had to have the produce of 1.5 acres to eat to provide the energy to pull the plough. At the time many learned papers were written on the economics of one against the other. The love of the horse was enough to sway the argument, together with the fact that a mare could be bred from. At the end of its working life a horse had a value, not that an Englishman ever said he ate horse flesh and I suspect a good deal was sold as beef. Labour was also still plentiful and cheap, it remained cheap up to WW1 but not so plentiful, for reasons that will be shown as we go along.

Apart from the use of horses to work the land, horsepower was also used to move everything in the country. The railways, yes, were in their heyday but the horse was still needed to haul to and from the railway station, and in fact their numbers increased due to the rise of the railways.
So it becomes clear that this animal was very much needed but was taking up a great deal of land required for human food production. Perhaps out of interest it might be as well to give an indication of the care a working horse had to have so that he was able to function at his best.

Firstly, he had to be given four good meals per day, well spaced out. His body had to be kept clean or he might become infected with nasty things that took the edge of his pulling power. All his excretions, deposited in the stable, had to be hauled away every day or his feet could become infected which would make him rather like a tractor with a flat tyre!

Although no work was done on Sunday that animal still had to be fed, cleaned and watered. A carter started his day at 5.30am, by putting fuel into the tank so to speak which took the horse about 1.5 hours to eat and digest. Work started at 7am out to the field. The team then stopped at about 9am for both carter and horse to have a snack and then work continued until 12 noon. Lunch until 1, and then work continued until 3.30pm. It was then back to the stable to be cooled off, cleaned, fed, watered and put to bed. The carter would always look into the stable at 10pm to give the horse hay and see all was well. So it was a long day – 6 days a week!

The routine changed on Sunday including the type of rations fed. If the carter happened to feed them the same amount of food, on a Sunday, as on a working day, he was asking for trouble. Because the working ration was so nutritious, a horse at rest would not burn it off and this would lead to one or all his legs swelling up. He would then not be able to work on a Monday morning as he'd blown a valve so to speak. And hence this affliction was called Monday Morning Disease.

It's so romantic to look back with nostalgia at the use of horses ploughing, but what an expensive prime mover they actually were!

With regards to Points 5 and 6 - recycling of fertility and the way forward, it has been shown before how plant foods e.g. Guano could be bought from around the world. This was the residue of the fish diet of the sea birds, dried over millions of years enabling it to be dug out of the ground and sold. The material contained high concentrations of both phosphate and potash, both needed by plants so that they could grow to their maximum. The other food which was required to complete the process was nitrogen. This element makes things grow fast and the only source then was from the legume family, clovers, peas and beans, which take nitrogen from the air via their roots, use it themselves and when they die or are ploughed under, release it from their decaying roots.

Not that any of this was realised at this time. The producers knew what material helped growth but not why. So in effect they were farming in the dark and could not make every move count. Today it would be called inefficiency. The story is very complicated, for although they were not putting the correct amounts of plant nutrients on, they were putting on masses of organic material or humus. This was in the form of dung, a mixture of straw and animal faeces, shoddy, a residue from the woollen mills, or soot and there was plenty of that - anything in fact that might help the soil structure. This humus is invaluable and without it nothing will grow. It has the ability to hold moisture, essential in the lighter soils, to open up clays and the heavy land and let them breathe. Cultivation becomes far easier in a humus enriched field. It can be said in general terms that humus keeps the soils body and soul together. The deserts of the world although lacking water or much rainfall have also lost their humus. As the rainfall declined so did the plants' growth and thus less dead vegetation, including the roots found its way to the soil. No doubt high temperatures played their part in speeding up the oxidation process. Once the fibre has gone, the wind can play havoc with soil particles, and thus you get these huge mountains of sand continually being eroded and shifted. As every good school boy knows, as the Teutonic plates move around the earth, deserts come and go due to the changes in weather patterns. This movement takes million of years to show any great change in the earth's appearance, so we have no need to worry about its effects on any of us. However, we might well have to think far more deeply about how we might harness the great deserts, to produce food or become rolling prairies where people might live more comfortably without all that sand flying about!

In the 1800s the landlords were men of great wealth. Many were not from the old landed gentry but had made money in the new industries springing up all over the country. Some had made money from sugar plantations via the slave trade and the sweet tooth of the population in England. Generally speaking these men brought a new approach to land management, how to make money was in their thoughts, and they had hungry townships developing like mad in the ever spreading industrial revolution. These fresh brains were eager to develop any improvements that meant cheaper and increased production of food. They were not afraid to exchange ideas, and farming clubs began to be founded all over the place.

An intelligent man was about to come on to the scene, a chap called Lawes. He was left a large farm in the county of Hertfordshire, at a place called Rothampstead. This was in 1843. It seems he was a chemist and was fully conversant with experimentation. Before he had the farm he was trying out chemicals on plants growing in pots. Now he could substitute pots for fields - he had the money and the knowhow. He laid out experiments that although primitive by today's standards have been continued up to this day on the same farm and one in particular in the same field. The foundations he laid in farming research still lurk at the back of today's research. Again the story makes fascinating reading, should you have been bitten by the farming bug, or indeed are interested in where your food supply is to come from in the future. Perhaps a small explanation on one of his research plots may give an idea of what he was seeking.

He took a field called "Broadbank" on his farm, then divided it into plots all of the same size. Each plot had a tile drain laid right along the middle, emptying into a ditch at the bottom of the field from where the drainage water could be collected, and thus kept separate. Cropping was with winter wheat, all plots with the same variety. One plot was not given any manure, while the others received manures of the day, such as I have already mentioned, but each plot had something different given it. As new products came to light they were tried out individually. The crop yield was recorded, and thus a comparison could be made, and any chemical coming through those drains was noted. As time went by a picture began to emerge showing which plant foods were needed.

The plot that had no manure of any sort given to it, is to this day giving a crop only yielding around 1 ton of wheat per acre. It appears that the plants are just managing to get a little phosphate and potash from the subsoil, as there are always these elements being pushed up from below. Nitrogen is another story. According to the drainage collected from this plot the amount varies each season, due to how much is leached by the winter rainfall. Potash and phosphate do not leach from the soil as does nitrogen which is soluble in water. These pointers have been of great help to the modern farmer, for he now considers the rainfall and temperature before deciding how much nitrogen to apply to his crop. He has learnt to use what nature has given or what has been left after the winter rains. That's progress.

After several years Mr Lawes took on a partner to help in his work, a chap called Gilbert. Between them they made a great discovery, perhaps the first chemical breakthrough in agriculture involving a chemical reaction. It needs a little explanation to understand what they were seeking. The phosphates they had were not very soluble in water, and thus plants were finding it difficult to get enough of the element to make proper growth. They were playing around with rock phosphate which can be found in many parts of the world and is a natural product laid down from past volcanic eruptions. Their concern was that if they put this rock phosphate on to the soil, the plants would find it difficult to absorb but if they could unlock the phosphate from the rock, the plants might well get more of the food. In the naturally acid soils this release takes place but, as early man had discovered, neither the wheat plants or in fact any farmed crops will grow well in acid soils.

Gilbert being a chemist soon found an acid that would do the trick. It was sulphuric acid, H_2SO_4. Now the results were so good, that the men patented the invention, and started a company to manufacture the fertiliser. They called the new product 'Super Phosphate', probably the first time the word 'super' came into the language. They made a great deal of money. I imagine the impact from this idea was akin to the period when the first axe was introduced, I bet that tribe made a killing, but they had to put up with food instead of cash. This meant, of course, that they could concentrate on manufacturing things and forget all about how to grow food.

As far as Lawes and Gilbert were concerned they wanted to see farming improve, so they put a £100,000 into a Trust fund which was to continue the experimental farm at Rothamstead, and spread knowledge on plant foods, and in time the control of pests affecting farm crops.

It's a funny thing, but the farming industry has always disseminated any new ideas that show improvements in crop or animal production. The agricultural fraternity rather like gardeners, want nothing better than to look at a bumper crop or perfect animal, and will tell whoever asks how they managed it. Other industries always seem to play their cards closer to their chests, better for the balance sheet I suppose.

Now is the time to take up the last point in the list of blips showing up on the Victorian radar screen, that is the need for National control of the Agricultural potential. This point arose I feel from the rather long and nasty financial depression that followed the victory in Europe over Napoleon. From about 1820 until the repeal of the Corn Laws in 1846 farming suffered badly. This was because of the low corn prices after the highs of almost 20 years of war. This led to slackness in the household budget. Money was borrowed freely, rents rose too much, remember that much of the land was tenanted, and more labour was kept than was needed. The adjustment was cruel. Farmers went bankrupt by the thousand, banks went into liquidation, landlords failed to let their farms. No doubt the better lands as always pulled through. By better I mean those on the top grade soils, or nearer to a good market such as London.

Farmers suffered, yes, but the farm labourer was devastated. New machines were coming on to the market to save labour, which was happening in other industries, not only farming. It was the era of Luddites, who smashed the machines that caused starvation. If you were caught stealing, even food to feed the family, it was off to Australia, which for some of the single chaps was the best thing to happen to them. All this is a story in itself which is well worth reading.

Parliamentary Committees were set up to find out what the problems were, and how to cope. It was around this time it was decided to set up a permanent Board of Agriculture, with a brief to find out what the country had in the way of agricultural assets e.g. acres, cattle, horses, and all the crops. It finally came into being in 1865 guided by Lord Palmerstone. To hone it into more or less what we have today i.e. DEFRA (The Department of Food and Rural Affairs, previously The Ministry of Agriculture, Fisheries and Food) took some 20 years, At last the idea had sunk in to the political mind, that farming had joined the scientific age, the age that had

been started in the 1800s. Farming could no longer be left to the yokels. Political brain power was needed to pull the strings and oil the wheels.

CHAPTER 9. THE CROPPING CYCLE OF THE NEW ERA

In my introductory chapters I attempted to show how farming might have evolved. It is a story that the archaeologists are forever unravelling. The progress they have made over the last 25 years has been tremendous, and the way technology is travelling, the next 25 years will push the curve of knowledge up at an even steeper angle. I make no apology for I am only trying to give a basic platform from which you may or may not want to take further. Should you not wish to read the subject in more detail, at least I hope you will be able to understand enough to make more sense of what may come along in the future.

The new era or dawning which had really begun in the 16th century showed how yields could be improved with the increase in fertility and with new crops and techniques, many originating on the continent. Now we are back to brains with names. The first of which was a chap called Jethro Tull. He was farming around 1730 and he probably had a private income so he could experiment without fear of insolvency. The average farmer at that time could only take up a new idea if it had been proved beyond doubt to work, so he tended to stick with proven methods. Thus things changed slowly. Tull was different, no doubt well-read and taken by the fashion for science being led by the Royal Society in London.

His thoughts led him to believe that if you could exclude weeds from the crops, they would have no need of fertiliser. How to do this was the problem. The seed corn at this time was broadcast (this means throwing by hand over the tilled soil as it had always been done). Weeds would then come with the corn, taking food and moisture away from the crop. A certain amount of hand-weeding could be undertaken but then the feet of the folk weeding tended to damage the crop.

The story goes that Tull, whilst sitting through a particularly boring sermon in church, let his mind wander amongst the organ pipes so to speak and was hit by a brilliant idea. If you could arrange to let the seed fall down the pipe the plants could only grow in rows and you had more chance to get at those weeds. Some have suggested that the sermon was about the parable of the sower, who knows? Thus we had the invention of the corn drill, followed quickly by the horse hoe which could go along the row and cut the weeds off at the roots. Poor Tull failed in his original idea of using the suppression of weeds as the way to greater yields without fertiliser, that did not work, but the drilling machine certainly sparked off others in new directions.

The good old turnip was introduced to England around the time of Tull by a man called Townsend. He had been travelling around Europe, as many men of means did in those days, looking at the farming in France and the Low countries. He came back with new plants to try out on a field scale. Besides the humble turnip he also brought over the red clover and Italian Rye grass. All of these crops were to make a huge impact on the countryside during the next 150 years.

So what did these gentlemen do which put guts into the soil, more and better bread on the table together with bigger and better meat supply for the rapidly growing population? Well they brought the opportunity of crop rotations into being. In essence this means one crop can help another by improving the fertility of the soil and keeping disease at bay, and as the years go by so the soil improves and the farmer get richer as his output grows. It was called at the time High Farming. Although the fire under the boiler was lit in mid-1700, steam was not reached until the mid-1800s.

Now we come to Coke who became the Earl of Leicester and took over his Aunt's estate in Norfolk. Of some 3,000 acres, most of it was scrub and light sandy soil. He too had been reading and travelling abroad. He took the knowledge he had gained and put it into practice on the estate. Many of his tenants had left their farms because of the impossible farming conditions, so Coke had the job of improving the soil, and then letting the farms back with a clause in the tenants' agreement setting down what rotation was to be used.

This began what was and still is called the Norfolk Four Course Rotation and it goes as follows: Barley sown in the spring is under-sown with a grass and clover seed mixture, the thinking behind this is that it is a way of saving time. If you wait until the next year or the spring to sow the grasses you will have lost a productive period. The clovers and grasses would not produce to their maximum in the year of sowing. So when sown with the barley, or a week or two after, they potter along at the base of the cereal but do not worry it. After the cereal harvest the young seeds begin to grow away. In the following spring they are in the mood to produce a great quantity of hay, almost certainly two cuts and even a crop of clover seed in the autumn.

Apart from all of this fodder, the clover roots have been busy taking all of the Nitrogen out of the air and storing it in the nodule on its roots, as illustrated in the picture below.

Now in the autumn of the following year this mixture is ploughed up and sown to winter wheat. Wheat is a hungry plant and all of this nitrogen combined with the humus rotting down from the mass of grass roots is just fine for it. After this hungry crop has gone, in year three, the field is semi-fallowed. That means that it is cultivated in the spring to make the weeds germinate and then they are killed by ploughing ready to sow the famous turnip in rows, about 18 inches apart, through which a horse can pull a hoe and men can also work a hand hoe, thus killing weeds.

Now Coke brought in the animal factor – sheep. These he folded on the turnips all through the winter. He fed them other things which were coming to hand at this time, such as Oil cake. All manure was being returned to the soil and thus creating more humus. The sheep grew well – better than they had when out on the old scrubland and were another crop to sell. So we are back to year four once again ready for the barley crop to be under-sown as before. The barley was in great demand for the brewing trade – another nice little earner that year.

This rotation was ideal for this part of Norfolk but not so great for the wetter parts or the heavy clay soils. Within a few years, the estate was transformed. Before Coke took over the estates farmers grew a little rye, reared a few animals and, as I have mentioned, as they were only scraping a meagre living, they paid little rent. It has to be understood that this part of England has now, as then, only about 20 inches of rainfall a year.

Another factor that played a big part in this reclamation was that of marling the soil. That is to dig clay from a pit and spread it. This helped the top soil to bind, hold the sand and get the whole process of holding the moisture on the road. Coke spared no expense and was well rewarded when, at last, wheat yields began to rise and at a time of rising prices. But I bet his greatest pleasure as a good countryman was to show others what could be done, and they came from far and wide at his invitation. He set a week aside once a year to hold demonstrations at his home, Holkham Hall and anybody who was anybody did their very best to go. For it was not only crops he worked on but also the breeding of better cattle and sheep.

The heavy land farmer had to use cattle as sheep did not thrive well on wet clay soils. For him it was a more expensive exercise by the fact that he had to have yards to keep the animals in during winter. This kept them off the heavy wet soils which would be ruined rather than helped by the cloven hoof. The goodly turnip had to be carted to the yards together with straw, the muck left behind had then to be hauled out to the fields and spread. However, these soils probably paid well because their fertility was inherently greater than those of Norfolk.

This simple rotation gave yet another plus to the farm which was that of disease control. Let me explain. Barley grown continuously year after year builds up diseases that pull down its potential. Clovers have the same problem. Turnips get a nasty disease called Club Root and wheat following wheat is subject to something called Take-all, which speaks for itself!

It must be remembered that around this time, land was being enclosed on a large scale. Of course it paid to do it and it must not be forgotten that it created a great deal of employment. As an example, surveyors were needed to lay out the land in suitable blocks. Ditches had to be dug and those hedges that people worry about so much today had to be planted and fenced for protection until the young hedge plants (quicks) had grown out of harm's way. New farms and houses had to be built and roads laid out. Many nurseries started up just to grow the hedging material i.e. the quicks and other hedging plants for they were wanted by the hundred thousand to create the landscape we see today. Good bye to the commons and the open fields. Coke had proved beyond a shadow of doubt that without this progress in the countryside, the nation could not be fed adequately. In the middle years of the 1800s, there was more land in cultivation than at the peak of production during World War 2.

Coke proved that his rotational system made soil that could produce good crops. He used all the aids that science had produced, but without the part played by animals, in his case sheep, and to a certain extent cattle, the improvement would have failed. The farm animals available in the mid 1700s were not the best when looked at from the point of view of the butcher, or for that matter the farmer, because most of the time they were short of food and almost starved during the winters. So as they could never show their full potential the butcher failed to get many good cuts out of the beast and the farmer found that his money was tied up for a long time before the animals were ready to sell. Before the commons were enclosed it would have probably paid because very little money was needed for rents and common land cost next to nothing. With enclosures, however, things changed, and money became very important. Landlords were looking for a return on their capital outlay and seed and extra stock needed financing. It must be realised that not all enclosures worked. It became fashionable to enclose because it was thought to lead to prosperity. However, land that was too poor, such as the upland moors soon reverted back to what it had always been – incapable of improvement at least during that period.

Almost certainly stock followed crops in the process of improvement. At last the swede and mangolds, harvested, stored, and then fed, not folded over (block grazed) as was the turnip, meant that winters were not the worry of the past. Indeed with the addition of a little residue from the Vegetable Oil Industry on which more later, cattle and summer lambs could be fattened during the winter on the new root crops. The new farming clubs that began to spring up all over the United Kingdom went out of their way to give good prizes to what were considered to be the best butchers' beasts. Breeders soon began to make money from their exertions, by hiring out bulls and rams to improve the existing herds and flocks.

The vast majority of animals needed for this ever expanding rotational farming came from the hill areas of the kingdom. This trade had been going on for centuries, always moving them off the hills onto the low ground before the winter began.

They grazed as they went and so increased in weight ready to be slaughtered as they reached the towns. Even today the trade in cattle and sheep runs into millions of pounds and has become more involved than it was in Coke's era. For instance, after a few years on the mountains, the teeth of a ewe become worn down. She gradually fails as she is unable to eat enough grass as it grows so low to the ground on the uplands. If she is moved to the better pastures of the valleys she may well live another couple of years and produce two more crops of lambs. The lowland flocks of today have a depreciation factor of some 20% and thus are in the market for replacement crossbred ewes from the hills. The mountain man, who has very little good land to grow crops for winter feed, has only then to worry about feeding his basic flock in the winter.

What a grand system had evolved by the 1850s. It worked – it kept up the food production for the ever expanding population. After Parliament got rid of the Corn Laws, which had looked after the farmer via the wheat price for years, and went for Free Trade, disaster did not fall upon the countryside. The price of wheat remained firm and more meat was being eaten. Things could not have been better and from about 1846 until the 1880s it was the golden age for farming. The riches did not only flow into the farming pocket, for the foundries that were increasingly busy building ships and railway engines and the rails, had not neglected agriculture.

New ploughs, cultivators, horse hoes for those turnips, reapers to replace the scythes, steam engines, threshing machines, buildings, more pumps, rollers – the list seems endless. As farmers made money, many ploughed much of it back into the land and improvement was the order of the day. Colleges to educate young farmers started up and seed and fertiliser companies increased to exploit the new seeds and plant foods. The practice of having farm workers lodged in the house began to end as cottages were built to house them. Unfortunately these were tied to the farm but they at least enabled workers to be more mobile and gave them the opportunity to improve themselves.

As machines began to dominate the scene, it became apparent that the new livestock was too valuable to be neglected. Those heavy horses that had been bred to pull up to 1 tonne at a time and kept the railways in full operation were ever in need of care. The horse was the fluid that kept life going and the new veterinary service which was, up until 1860, rather along the lines of an African Witchdoctor, turned professional. To keep all this revolution in line for the national good was that Board of Agriculture who knew what was afoot in the countryside and could keep Parliament properly informed.

This proper footing did not come about until 1889 but in a way this was only putting under one roof what was being done by other ministries on behalf of farming. 1865 was the turning point in the final decision to bring about the new ministry. This was the year that a most vicious cattle disease came into the country and the Veterinary profession had no way of combating it apart from mass slaughter. This disease was Rinderpest and it required legal action to enforce and this was not forthcoming.

The farmer was ignorant of the contagious nature of the disease and so did not understand or feel the need to cooperate with mass slaughter. He was ever hopeful that his animals would recover. If he wanted to sell animals from his farm he could see no reason why he shouldn't be able to do this. If it spread the problem he was not concerned.

The losses from Rinderpest were running into millions of pounds and the government was worried. The new improved breeds were indeed stocking the world including the new colonies and apart from this home disaster the disease was likely to affect the export trade. So in 1869 after some legal hiccups a slaughter policy was introduced and isolation became compulsory. Compensation was given and the problem was solved.

This policy was administered by the new vets and it also covered other diseases such as Foot and Mouth and Rabies. Imported animals were quarantined or inspected at every port. England became an Island fortress as far as livestock was concerned and has remained so.

The golden age of farming was also the golden age for the whole country. Not that those who worked in the new factories would agree. Nor would those who had to resort to the workhouse for assistance be in any hurry to condone the system. It must be remembered that the country was undergoing the biggest upheaval in its history. I doubt that anything had been so dramatic since the Romans had left.

The drift to the towns had been gathering pace right from the 1700s and continued almost up to the 1980s. The standard of living was increasing no doubt albeit in a patchy way. Speculators were making big money building new towns, factories and the infrastructure that goes with the system.

To me a sadness was there, mixed in with the new life. In the past with the open fields and the common lands, small rural industries such as weaving and metal work all had rural connections. Everyone had a hand in growing their food, access to the common for their cows, for wood and even for berries. It seemed that everyone helped each other with the seasonal work such as haymaking and harvesting. This sounds idyllic but no doubt it was hard graft but it seemed to have bred an independent spirit. It set the mould in the English character of fair play, the recognition of evolution rather than revolution and justice.

But the commons went and the factories came and things changed. British farming was in a new era. No more strip farming – if a man could save a little capital then he might well get on the embryonic farming ladder but it was more like snakes and ladders in the early stages.

Strange as it may seem there were more tenant farmers around than owner-occupiers. This was probably because the massive enclosure movement meant that the big landowners could gain far more from rents than they could from farming the land themselves. If they wanted to take a farm back in hand they could do so with ease. Tenants only had a yearly agreement and in most cases it was renewed again each year, thus the tenant could farm away quite happily improving the land, breeding better livestock and almost certainly putting up buildings on his own account.

However a tenant was often at the mercy of the landowner and very often great hardship was caused when he was evicted from his farm. The tenant had no legal rights. It may seem that he had nothing to lose other than his way of life but when one looks at the implications he could well lose his capital and be in no position to take on another farm.

Over hundreds of years a voluntary system of compensation had been worked out in certain parts of the country. Ireland was one of the first places to try and make sense of the whole thing but being a voluntary arrangement it was abused by the landlord. Mind you a tenant could also be at fault by exploiting the soil then leaving behind an exhausted farm that the landlord had to spend money on before it could be re-let or he could farm it himself. Coke's story illustrates this very point.

A simple example may show the sort of thing that used to happen. A farmer taking over a new farm usually puts a great deal of time and effort into improving the soil.

His agreed rent is usually fixed on the inherent fertility of the soil, so if he can push up yields it is to his advantage. Of course, he does this by putting more and more dung onto the land, buying in fertiliser, putting on chalk to kill off the acidity, putting up simple buildings to house stock along with fencing and drainage. He will have sown grass and clover crops that will have occupied the field for 2 years or have winter wheat crops that take nearly a year to mature. Now, before 1880 he could have had a notice to quit with no compensation for the improved fertility, the crops or the buildings. Of course a tenant could be evicted for failure to pay his rent but equally he could not get his Landlord to repair or renew buildings.

From around 1880, Parliament started to wrestle with this problem which went on for many years. They were hampered by the fact that any laws passed were blocked by the House of Lords many of whose members were of course the landowners in question. However eventually 1903 brought a pretty good Act into being.

This Act of 1903 laid down the responsibilities for both tenant and landlord. This included what the tenant could or could not do to certain fields and also what forage such as hay and straw he could not sell unless he fertilised the ground to replace the lost fertility. All good stuff which had come about with the increased knowledge of how the soil works.

An arbitration system was brought into play which could be used if the farmer and tenant could not agree. This did not happen very often though, for valuers acting for both sides nearly always reached an agreement on a give and take basis.

It is interesting to note that Surveyors and Valuers really came into their own as the high farming got going. In fact as the Enclosures movement was in its early stages, anyone who had a good grasp of maths was in great demand. Fields had to be laid out and measured to the nearest inch, rod, pole or perch to determine ownership. The tenancy agreements became more complicated with the calculations required to determine the pricing of unexhausted manures, crop values, ricks of hay etc. Valuers were in great demand which created a whole new professional group called Estate agents. They in turn extended their activity to auctioning farm produce in the market towns. The old system of farmer bargaining with farmer or customer was on the way out.

CHAPTER 10. THE DISCOVERY OF THE NEW LARDERS

It took many years from the discovery of the new lands to their exploitation. Like the Industrial revolution, that really began in the mid 1700s but was not evident until many years later. The Americas were discovered in the 1500s, but Australia and New Zealand not until the mid 1700s.
Looking for new lands which would be useful for food production did not enter the heads of any explorer during this early period. The biggest influence on farming came in a quite unexpected way. It was the rat with the flea, all the way from China which did the damage. As every school child knows, these rats brought the Plague, or Black Death as it was called, in 1349. Most of the damage was done in the 14th century but it had a nasty habit of coming back over the next 150 years. There was no cure and if you got it you were dead in a matter of days. Thus in a matter of months the population was cut by half. In my mind there is no doubt that the outcome of this was a turning point in the way English farming was to develop.

Lords of the Manor, who at this time held sway over all land activities during the strip farming era and the feudal system, suddenly found life very difficult that is if they themselves did not succumb to the Black Death. The Lord could call upon his serfs to come and work his farms before they could go and work on their own patch. Life was hard for the dogsbody serf. He was unable to make any economic progress of his own.

When those that were left after the plague came to their senses they soon found that they were in great demand by the Lord of the Manor who was desperate to get his field work done. Up went the wages or rather 'time' was converted into money. However this was not really the answer. So the Lord started another system that of 'copyholding' or letting his farmlands out to serfs in exchange for a rent or payment in kind. The serf had the land for as many years as his family had their names on the court roll, or copyhold. So they could stay put until the last one faded away. Rather a long-term tenancy agreement which was not to return until the 1970s.

Such was the devastation of the Black Death that even this did not work very well and again the Lord and Master had to come up with a new idea. This he did when the demand became so great for English wool for export to the continent where it was in great demand to make fine cloth.
The 'death' was, in this case, a godsend. Areas of land had lost whole villages – they had become ghost towns. Wages were high and labourers still in shortage so the Landlord turned to dog and stick farming. That is he turned his domain into a vast sheep walk. He had no need to have any close shepherding. If the sheep could find enough to eat on the deserted open fields that was good. If some died it was no matter. They just needed to be gathered once a year, sheared and the cash rolled in. It was not long before they found that they could add value to the wool by weaving it into cloth and this was undertaken more and more as the population started to increase.

Even the diminishment of Lords of the Manor from the Black Death allowed any survivors, be they other Lords or even serfs, could take over the lands and add them to their own. There are families in the country today whose wealth goes back to this period. Those serfs that had managed to get land could at last begin to build up capital setting them on the road to becoming what is known as a Yeoman Farmer.

It was 300 years before the other great change was to take place which was in large measure due to over population and religious persecution. The Spanish were the first to explore but not with food in mind. They were after gold and spices. They used the excuse that they wanted to convert any heathens they found to Christianity and went westward looking for these things. Spices were of course in great demand because they made palatable the rather dull diet of the times. Or it may be that they made people smell better after a winter of living in woollen and leather clothes, no doubt it was needed!

Now the British, after they became tired of pirating the poor old Spaniard, looked to North America. First off the good fishing lured them to Newfoundland so I suppose you could say that it was food driven exploration. They explored the east coast of America where they picked up the bad habit of smoking and then exported that home to the population who also became addicted to the tobacco leaf. That was a money-spinner and has been ever since. Then on to the West Indies where the sugar cane was grown and the industry expanded massively with the use of slave labour. Once again for many British this proved to be a huge wealth creator. Large estates were purchased in the home country with the money generated and this enabled the seeds of 'high farming' to be sown. Still today you can find several families connected to sugar production combined now with chocolate.
None of this could have taken place without the use of slaves from Africa, but it had no direct effect on farming in the home country. It was the colonial expansion that went with this era, keeping out the Dutch and the French and having the freedom of the seas to trade throughout the world. The luxuries from the east were what we wanted and these riches had nothing to do with the basic foods.

To round off this little bit of potted history, we must include the acquisition of Australia and New Zealand to what became the British Empire. Captain Cook was the great seaman who did the business so to speak. The lands he found were really

incidental to the project he was set which was really to do with the navigation of the sea, again linked to trade. The great areas he found never looked to be of much use, until the Americans chucked us out of that colony and we had nowhere to send our criminals - Australia seemed to be the very place to send them.
Within 8-12 years the interior had been checked out, the potential realised, and men put on the spot to do the job. Add to this a bit of science and you had the ingredients to topple that high farming of the Victorian farmers.

After the long war over the slave trade and the expansion of the railways, combined with a mass movement of people, a dagger was about to be plunged into the heart of British farming. In fact I think you can probably say, three daggers were about to be thrust into the heart of British farming.

The first dagger was the decision to become a free trading nation back in 1846 allowing the price of wheat and therefore bread to be at the mercy of the world market price where previously it had been protected. For a while nothing changed, although the debate was long and hard and the landed gentry fought tooth and nail in opposing it. But the new men of industry were in the ascendancy and won the day. At the time apparently, Disraeli told everyone that farming would be devastated and for many years he was proved wrong. However when he at last became Prime Minister his predictions came about and with a vengeance. By this time he was an old man and anyway his mind was on other things. He did not have the will or the political backing to help farming and it was sold out to the industrial base. They sold the New World their machines in cxchange for the cheapest food they could get. This ensured that wages could be kept low.
The second dagger came flying in from the U.S.A. but before I go into this I need to give you a bit of background.

The Americans had not been idle during the post revolution period and they began to look at the Great Plains that they had on their doorstep. After the American Civil war the race to bridge the gap between the west and east coast of the USA began. The success of the railway system in winning the war for the north influenced the government into subsidising the building of a rail link across the Great Plains, and the mountain ranges to California. They started from both ends at the same time and after three years they joined up more or less in the middle at Promontory, Utah.

Now the rail link was complete it needed trade to make its builders money although the rail men had 'milked' the government all the way, in a scandalous fashion. Huge tracts of land on each side of the railroad were given to companies, upon which they were to encourage farmers to settle. Strings were attached – the farmers were expected to send their produce on the new railway at whatever charge the company wanted to make.

The story of the settling of the Middle America is well worth reading. The fight that took place by folk, who were sold a 'pup' by these railroaders, eventually formed the creation of the Democratic Party of that country.

From 1870 onwards great efforts were made to get people from Europe to become immigrants. Meetings were held in nearly every continental country and before long folk began to pour in. When one realises that the acreage that was there to be farmed, before even needing to think about irrigation, it was a godsend for the landless men of the Old World. To further encourage this steady trickle of humanity, the American Government gave free blocks of land of around 160 acres to those that came along.

At first the new farmers found life extremely difficult upon the Great Plains. No trees, hellish winters and very hot summers. The old prairie turf was found almost impossible to cultivate. Having come so far these folk, many now without any money, had no alternative but to stay and do their best. If it had not been for the dried dung, left by the fast diminishing Buffalo, which could be burnt, they could not have survived the early winters. As it was they had to build houses out of the turf, besides digging a hole in the ground to keep out of the everlasting wind.

The iron shod ploughs with a fixed coulter or knife to cut the sod would not work as the fibres in the ground, which had developed over thousands of years, were like thick coconut matting. The slice, if you did manage to cut it from the main block, would not slide up and over the turn furrow. An observant chap, or should we say Brain, came to the rescue. He was repairing a plough one day, the turn furrow in fact, and was using a metal that had once been part of a circular saw - high tensile stuff. Lo and behold the turf slid over it well and combining this with a circular coulter, rather like a wheel, the turf succumbed to the plough. That was the catalyst that unlocked the pent-up fertility.

It was Germans from Russia that came along next. They had been lured to Russia back in Catherine the Great's time when Russia was looking to improve or at least catch up with its neighbours. They had been wanted because of their farming expertise and thus show the way forward to the peasants of the Volga region.

With the coming of the Amcrican incentives i.e. 160 acres, these Germans realised that the new lands of America were very similar to those they had left i.e. extreme climatic conditions but with the potential to make a good living, and a freer life. How many men came I have no idea, but each carried with him a bushel of wheat (about 62 lbs. or 30 kilos). They had been told that wheat was not a crop that did

well on the plains but they must have had a great faith in the variety of wheat they had bred to have heaved the seed half way around the world.
It was called Turkey Red and it loved the prairie soil and withstood the climate to produce better yields of wheat than any other variety that had come from wetter Europe.

As I have mentioned before, good news travels fast in the farming world. It was certainly the case in the States for within the space of 15 years, the state of Kansas was the biggest wheat producer in the world. It was also the cheapest wheat producer in the world thanks to the pent up fertility within the prairie sod. No high farming was needed to achieve good results. No expensive rotations came into play and as labour was not plentiful every chance was taken to use the latest mechanical breakthrough that the industry had to offer. Still the biggest problem was the climate, for just at harvest time the plains' weather could provide devastating hail storms that could strip away the ripe heads. Many famers were ruined but many more prospered.

The United States was not the only country at work on the plains. Canada was going the same way. The Scottish immigrants, who had flocked to the eastern Maritime Provinces back in the 1850's, were now on the move westwards as the railways began to open up the country. It was an even harsher environment than that in the States, but the toughness of the immigrants more than matched that of the climate. It was in no time at all that they too had developed a way to grow wheat without rotations.

Back to our second dagger.
With the coming of the steam ship and with great waterways in both these lands, the cost of shipping wheat was coming down each year and by 1875 the first load

arrived in Liverpool docks. The landing of the first wheat from abroad also coincided with some of the worst weather for farming in England. Wet harvest followed wet harvest – this after 30 years of good farming weather.

By 1885 the acreage of wheat grown in the UK had dropped by two million acres, something like one millions tons of grain. The British farmer simply could not compete on price. His high farming was very good for the soil and he was probably the most scientific husbandry man in the world but the price of wheat and therefore bread was going down each year to the benefit of the average man in the street. He and the new political masters could still remember the hungry 1840's when the price of bread rose almost out of reach of the man on low wages. No more of this. A full belly meant a more or less contented nation. If the farmers did not like the status quo in England then they could always go to the New Lands and start afresh – with no forelock tugging to a Landlord required!

Besides the loss of 2 million acres of winter wheat in Britain, agricultural wages began to slide, not that they were ever as good as those obtainable in the towns. The result of this was that 100,000 workers left the land. They had no real trouble finding work for they were soon employed producing material for the New World in exchange for cheap food.

In comparison, in the 1860/70's, the governments of European States put a tariff on any food imported from the New World. The reason for this was to protect the fabric of the countryside and the social structure. Their populations were so much nearer to the magic of the land and growing crops than was the case in Britain and of course one has to remember that they had so much more good arable land than we did, together with a much lower population.

Another point that must have arisen concerned the millers. Because of the climate, English wheat was always sold with a higher moisture content than imported wheat and the millers could thus add more water to the imported wheat flour, and make more profit. When Canadian wheat started to come in they soon found that the high protein in that crop due to the fast growing conditions in Canada, made the bread rise better and it would keep fresh for more than one day.

Of the 2 million acres gone from the wheat acreage – what replaced it? The short answer was nothing, it fell back to grass but with no extra animals to utilise it. Within a few years the farmers in Britain were in serious trouble. They had assumed that the farming system that they had evolved over the years would continue for ever. There was no doubt that it was a sound system and that people came from all over the world to learn its secrets. Yields were good and sustainable and harvest weather was the only fly in the ointment. Up to this time farmers found no need to cut labour but when wheat was being landed in increasing amounts, ruin looked them in the face. Labour on the farm had to be shed and now they had to think about improving output with far fewer men. The US prairie farmer was already using the tie-binder whilst the Scythe man, followed by his family making the sheaves, was still the norm on English farms.

Because the profits had been good year in year out, the banks were eager to lend money to the farming community and that community was only too eager to take it to expand their farms or to improve them by drainage, marling, or massive use of farmyard manure. Besides all of these, the temptation was also to improve the farmhouse, educate the children and generally show off their wealth. So when the crunch came and the banks wanted their money back, things went from bad to worse.

This bit of economic 'blowout' had happened before, after the 1815 victory at Waterloo. However, too much soil had passed over the turn furrow for the new generation to remember that time. In life, as we all know, it is easy to forget the bad times and only remember the good.

In a matter of some 10-15 years, from 1875-1900, British farming was in a devastated state. Only the very best of land was holding its own. Much of the rest was abandoned or left derelict – a sort of enforced set-aside as introduced in the 1990's. If it had not been for the rich men coming out of industry with a mass of money and wanting an estate on which to rear pheasants, life would have been a lot more difficult.

For all the time that the Navy wanted good wooden built ships, landowners looked after their woodlands as it paid to do so. With iron-clad ships now coming along, the oaks were not in demand and building timber could be brought in from the States and Canada for very little money so the building boom of the times did not want British timber. The result was a decline in English woodlands that continued until the Great War (World War 1) came along. Farming had to change gear but how?

So we come to the third dagger. This dagger plunged into the heart of English Agriculture was the invention of refrigeration. Scientific knowledge gained during the 1700s and early 1800s was being exploited in the form of a practical application to everyday living.

Australia was the dagger man this time. Those merino sheep that had taken the eye of King George, loved the Australian conditions. Their wool was keeping the factories in Bradford humming. But settlers from home had at last realised the grazing potential of the vast open grasslands they had discovered beyond the Blue

Mountains of New South Wales. The Australian government had developed and encouraged the owning and leasing of land. The country was no longer just the place to dig for gold. New Zealand soon followed suit and before long the sending of meat to the devouring hoards of the Industrial revolution was supplemented with cheap butter to go on that Canadian bread.

Money was also pouring into Argentina, to build railways and to gain access to the Great Plains or Pampas. To pay for all of this they in turn had to send beef to England. So fertility was being taken from the New World and poured into the population which was soon sending it out to sea via the sewage works.
Yet another little knife was also nicking the flesh of farming – it was the use of vegetable oils to make margarine which was a cheaper fat than butter to spread on the bread. These oils were used to make soap in the first instance, a material much in demand due to the increasing medical knowledge brought about by Lister and Pasteur. In the crudest terms the cleaner the population became the bigger grew the national appetite.

The use of these oils was a double-edged sword for British Agriculture. Margarine reduced the consumption of butter and so impacted on the Dairy industry but on the other hand the by-products such as cotton seed oil, palm kernel husks, and others made good food for animals and were cheap.

A way out of these new challenges was to go for the food fresh market. And so it began to become a case of 'Up Horn and Down Corn'.

CHAPTER 11. BRITISH FARMING CHANGES COURSE ON THE ECONOMIC HIGHWAY

Let us stop and have a quick look at what stage farming had reached before it had to change gear, downwards unfortunately, to cope with the new economic phase which was upon it.

The high farming system was good farming in anyone's book in thc 1890s. Rotations, recycling fertility with additional mineral manures bought in and a livestock system that had led to the improvement of stock breeding for the market place. The cow had to wait before it moved from a dual breed i.e. used for both beef and milk, into solely producing milk. The horse however was brought to its peak around this time. The Shire, Clydesdale and Suffolk Punch breeds moved almost everything to and from the railways. They could move great loads on the improved roads as well as plough the fields and scatter or sow the seeds.

Their management had also reached a very high standard, due in no small measure to the professional veterinary service that had come about in the farming boom time. There had always been folk with a certain expertise in the treatment of animals, similar in a way to the 'African medicine man', but as people began to understand the new science in veterinary work their days were numbered.

Mechanisation was on its way albeit slowly but mobile threshing machines were now becoming common, the flail and the Luddites long forgotten. On all the well managed farms the self-tying binder had beaten the scythe and sickle back into the barn for the first time since long before Roman times.

A Board of Agriculture was well established, its teeth only showing when the eradication of dreaded diseases needed the backing of the law. Its other function at this time was to keep a running record of all of the crops that were being grown in the United Kingdom, whether grass, cereals or roots. This recording started about 1880 and has become known as the 'June 4th Returns'. At first they were not compulsory but still gave an idea of the status quo. In time they were made the law of the land and could not be evaded. Soon all livestock numbers also had to be recorded and thus the state of the industry was always known and up to date.
Research stations were also established and farm colleges on the increase to keep up with any new developments that came along.

Now we must move along and see how farmers coped with the prairie shock and the imports of cheap beef.

The British Beef breeders had gone to town on beef improvement and had made money out of the export of different breeds of bulls to the new lands. Different breeds for different climates. For example Shorthorns and Sussex went to Australia as they could deal with the heat.

One of the biggest exports was the Hereford breed to America. This was an animal bred to convert fodder and grass into flesh. It is hardy outdoor type and was snapped up by the Americans in the first instance, to improve their Texas Longhorn that the Spanish had left behind in Mexico and Texas. Those cattle had no fleshing qualities but the new railways were ready to take beef from the ranges of the southern states to the new affluent people in and around Chicago. The Hereford transformed these herds and made Chicago the meat capital of the world.

Hereford Bull

This American boom kept going until about 1890 when the weather changed for the worse. A series of very cold and snowy winters together with overstocking of the ranges stopped the profits dead in their tracks and a more sensible policy came along with the conservation of fodder for winter use.

Aberdeen Angus Bull

The Australians were also after good sires, for they had none of their own, but they went more for the Beef Shorthorn and the Sussex which seemed to do well in the hot areas of the Northern Territory. However, the breeders never faltered in their overseas sales. Bulls were needed year in and year out to improve and service the herds abroad. Only the fattener on English farms found year on year that the cheap beef was cutting into his market.
In a way you could say that the British farmer's downfall was hoist by his own bulls!

It has to be remembered that the working population, in the UK was now in a financial position to buy more red meat. It was no longer just within the reach of the Lord and Master.
'Up Horn and Down Corn' by 1900 was the gear change that took farming along if not forward. The beef boys were changing their system to compete. Refrigeration

was getting the meat in but it was not of the best quality and was certainly not in any way fresh.

The root crops were becoming relatively dear to produce but wheat, together with oats, barley and by-products of the soap industry such as Palm Kernal cake were very cheap, and, along with other feeds such as linseed and cotton cake which were very palatable to livestock, large quantities were soon being fed. This increased the dry matter intake of the cattle and thus the speed of fattening.

Fresh meat commanded a better price than imported so a niche in the market was found and the turnover of the farmer improved. Sheep too had undergone a similar transformation for again fresh lamb and mutton was in good demand and breeds were undergoing a change to smaller and meatier carcasses with breeds such as the South Down coming along to fill this need.

Because of the cheap imports of wheat the pig was also enjoying a new lease of life. So the public could now have a good choice of meats and not many families went without their Sunday joint.

The human population was still growing, the very best soils were managing to make a living for their farmers from wheat growing but the poor ground such as that found in Norfolk (Coke country) was going back to nature and the rabbit from which Coke had rescued it. The hidden assets were accumulating in the soil from the imported fertility from around the world, but no one at this stage has any use for it. However the day was coming when it would be needed.

Around this period the great milk industry began. This was another feather in the economic cap of the country, but it took some time to hone it into shape. Farmers were initially very reluctant to milk cows. It was, and still is, a twice a day job and thought to be best left to the small holder and his wife or those farming the grass counties near London such as Buckinghamshire or the great cheese county of Somerset, not the good arable districts or the sheep runs. In time those views did have to change or the farmers would go under. It was only the hiccup of the Great War that temporarily slowed the trend into Dairy cows.

Until the arrival of the railways and for that matter for some time after, the needs of the population as far as milk was concerned were supplied by the town dairies. Certainly London had no other choice, but the smaller towns would have access to the country and the chance to buy fresh milk from there. It becomes clear as one reads the story of farming that most small farmers would have kept a cow for their own use, and if they were near a town the chance to sell to those who were unable to do this was too good to miss.

Because milk will not keep for many hours, out in the rural districts butter and cheese were the only way of turning it into cash. Butter was traded every market day in the nearest market town but cheese was traded in the autumn at cheese fairs, held at strategic points all over the country. This was due to the fact that hardly any milk was produced in the winter months, simply because there was no good food on which to feed the cow. It was only grass in the spring and summer that could do the trick and the only predominant producers were in the grass growing areas of Somerset, Cheshire and Buckinghamshire.

As cheese needs time to mature the autumn was the best time to sell and the only chance the large towns had of having a dairy product that would keep for a very long time. No doubt butter was salted and put into containers and with a bit of luck would keep for six months before becoming too rancid to eat.

That was the way for hundreds of years until London became too big to have any good access to the countryside and fresh milk. So the gap was then filled by the town dairyman. The forerunner of the big distribution companies of today. It seemed that almost every street had a chap who carried on this trade. The practicalities of doing this job were not that simple, because everything on which the cows were fed had to be brought into the town and everything that they evacuated had to be taken away. Although this manure was no doubt snapped up by the market gardeners that lived within carting distance.

What seemed to have become the normal system was to keep the animals down in the basement and store the fodder such as hay on the floor above. The hand work was quite considerable but no doubt the price of milk or the demand was great enough to make it pay well. The last London cow keeper only went out of production as late as the 1950s.
As it was almost impossible to keep a bull in the town to service the cows, the cows had to be brought from outside already calved, and tied by the neck in the cellar. They would have been milked, probably for about a year, until they dried off which was and were then fattened up and sold to a local butcher.

As the railways began to radiate from the capital city like the spokes of a wheel, so things changed. The fact that the speed of the trains running day and night was greater than the speed the bacteria could turn the milk sour helped. London, and the larger towns and cities, were becoming very affluent and had spare cash to spend. Because of this they could afford a better style of living and that included eating and drinking more dairy products.

Dairying was a diversification from the growing of wheat and due to the increase in the railways from 1850 to 1880, and a downturn in the price of wheat, dairying became a major industry although it took some 30 years.
At this stage the ability to feed cows a good milk-producing diet in the form of hay, roots and oilseed cake, meant that farmers could calve some cows down in the autumn and keep this valuable new market supplied all year. Summer production was however still great, for there was nothing like fresh grass to make a cow produce milk. The surplus was turned into cheese and butter but the value of this was suffering by 1900 from increasing competition from the colonies, especially New Zealand. They had the great advantage of a climate that could grow grass almost all the year at really no cost.

However the 40 million people of the British Isles were taking more and more milk each year into their diet in the liquid form. If they only increased their demand by 1//2 pint each a year it meant the need for another 5-6,000 cows to produce it.

It was not long before the good corn growing farms away from a good grass growing climate and with no piped water supply, started to become cheaper to rent. In many cases in the poor thin soils of Norfolk they were impossible to let. As always low rents and labour costs offered opportunities for some men to go into milk production. New cowsheds were built along with a cottage to house the milkers, wind pumps built to get water from wells and money was to be made.

The milk was sent in many cases both night and morning by train. It was conveyed in 17 gallon churns to the company in the town who had a fleet of ponies and traps going round three times a day to service the customer – pouring the milk from the churn into the waiting jug. As refrigeration had not become part of the house furnishings, the whole system demanded continuous supply and as there was plenty of labour about this could easily be done. The cash flow of the producer was improved beyond measure. He got a cheque each month, not once a year as with cheese production, thus his credit with the bank became so much better, making expansion so much easier.

However the system was without any physical controls and was open to abuse. Milk could be watered down at the farm – as it could at the town dairy. The farmer could be short measured at the depot. Who was to see fair play? The milk could be declared sour at the point of delivery and how could he argue?

Government was beginning to take an interest in the welfare of its citizens and so in about 1895 they set a standard for milk, below which it could not be sold. Not a very easy thing to police, but it was a start of the control of a fine food that was so vital to the health of the ever increasing population.

The other spin off from all of this change starting in the 1800s and the blowing ill-wind from the overseas farmer was the rising influence of the Agricultural Merchant. He bought and sold seed of both cereals and grass, manures were in ever increasing demand until 1880 when he found a way to make up this loss by finding and supplying cakes and cheap grains to feed the cattle on.

Mechanisation slowed up after the corn boom but exports of labour saving machinery more than made up for this loss. The shipping industry was booming with all of this activity, especially the bringing of the fertility in one form or another from abroad. But 1914 brought a war that was to change the gear once again although only for a short time.

CHAPTER 12. WORLD WAR 1 AND FOOD

The reason why the war came about would require another book but the sad fact was that it happened and all it proved was that effective weapons could be made to kill mankind.
The American Civil war of course paved the way in this so called art but 45 years after that great event no one had learned the lesson. After 1918 and the end of the hostilities the people called it the 'War to end all Wars'. It was not.

The bitter peace that overwhelmed the Germans ensured that they tried war once more. Their reasoning was the need for more space and the room to grow food for an ever expanding population. 80 million people take a great deal of feeding.

Britain failed to learn anything for future use, or so it seemed for some years, but the seeds that were sown then in a few men's minds germinated by 1939 and just managed to save the world from a mad man's domination, but it was a close call.
The general attitude when war was declared in 1914 was that it would be over by Christmas, the Germans would be put back in their box, and life would continue in peace as it had done since the defeat of the French In 1815.

No one thought of the possibility of the cutting off of supplies to the UK. We had the best navy in the world, we had kept the seas free for countries to trade, and it was thought that there was no reason why the Germans could or would be any worry. Up to 1915 when it became obvious that the contestants were at a stalemate militarily, trade was hardly interrupted by the sinking of Merchant ships.

Now the enemy began to see a way out of the stalemate of trench warfare in France. They began to sink more ships, not only in British coastal waters, but also further afield. These ships not only carried war material but food from the larders of the world to the UK, the habit of which had been growing since 1875 as explained in the last chapter.

By 1916 the English larder was looking very bare and something had to be done pretty quickly or 40 million people would starve, may well revolt and call an end to the war. The navy reacted by organising a convoy system for merchant ships so that they could be protected in groups. This took time to bring into operation as the independent merchant man was not used to being told what to do and stern measures had to be employed to get them to come into line.

This big war demanded a massive amount of raw materials which required paying for. Thus there was a conflict – money for food or money for war materials and as it happened the price of wheat in Canada and the U.S.A went up making the problem worse. It did not go up totally as a reaction to the war, as it often does, but because a nasty bug had decided to roar into those intensive wheat lands of the New World and cut the yields way down.

Mono cropping on as large a scale as the Americans were doing had never happened before. So a build-up of disease was bound to happen. One could say that the French had experienced something similar in their grapes in the form of mildew which they overcame with the chemical spray called Bordeaux mixture. The disease this time was Black Rust which fed off the leaves of the wheat plant, reducing its ability to fill the grain. At that time it was impossible to spray such a large acreage even if an effective chemical was available.

The problem was overcome by discovering the life history of the Black Rust. As it happened this rust was similar to Brown Rust which needs another plant on which to overwinter. If this plant could be found and eliminated all would be well. The shrub was soon identified and destroyed over huge areas and with it the threat to the spring wheat.

Back to the war. Although the price of all farm produce rose, farmers were reluctant to get ploughing for several reasons. Many could still remember the slump in the arable sector. The chaps farming at the time had fathers who were ruined or who got out of arable and went in other directions to survive. Also the thought was that the war would be over, if not by Christmas 1914, at least by Christmas 1916, so what was the point in getting geared up to grow wheat? To do this would need capital to finance it. Seeds, fertilisers, hedges and drainage would need to be reinstated and extra horses and ploughs to be bought. If the war ended suddenly and the world went back to 1914 economics then there would be more ruin and despair!

To sum up, thinking in modern day parlance, "no way Hosé." The war was being lost, so the government bought out big sticks and carrots to try and get extra food grown. They appointed local farmers onto district committees with the job of making the farmers get ploughing. In addition the prices of crops were guaranteed to the farmer for the foreseeable future.
Each farmer was made to plough up part of his farm and he had no option other than going to prison, and this, combined with the great prices offered, gradually began to get results.

Of course other factors were brought into play. Labour was becoming short as so many men were joining the forces. The factories were hungry for workers to make the munitions and ship builders couldn't keep up with replacing the shipping losses. Without food none of these products were worth a penny so, somehow, the land had to become the fourth line of defence. Ford tractors from the U.S.A. were imported to help replace the horses that were being used and killed in large numbers on the western front. Gradually the wheat acreage began to reach that of 1889 and the potato acreage was greatly increased. The cry of 1885 was reversed to 'Up Corn Down Horn' for more food per acre could be gained from wheat than cattle,. The price of land began to rise for the first time in years and the tenants were starting to look to buy their farms where possible.

Wars demand a great deal of money so taxes have to rise one way or another. Death duties were a way of doing this. On death the assets of the dead person if they were over a certain limit, were subject to a great percentage of tax on the estate. In almost all cases it meant that the estate, whether in stocks and shares or land, had to be sold to pay out the government. Some of the wealthiest men were being killed on the battle front and so the exchequer was doing well. If the estate was in land, as it was more often than not, it meant a great deal of land coming onto the market. The tenants, who now felt that times had changed for the better, were keen to buy and to my mind this era marked the beginning of the end of the tenant- landlord system.
Like all things in farming the process took some time to change and it was not until the 1990s that owner-occupiers outnumbered tenants.

So farming had suddenly become the fourth line of defence. Without food no one could work or fight and by early 1917 it was estimated that the country had only 3 weeks supply of food left in stock.

Around this time the numbers of soldiers slaughtered on the battlefield was greater than the number of volunteers coming forward, so conscription was brought in to

try and fill the deficit. Farming then became a reserved occupation as men were too important to be taken from the land; their skills could not be wasted fighting when the production of food was so vital.
Ploughing, sowing and reaping were not glamorous and there was no uniform but it saved the nation from semi if not full starvation. Food production was still very labour intensive and to help out with this the Women's Land Army was started, and although at first they were laughed at, many thousands came onto the land to replace the men.

So from 1917, the gear change in production was rapid from bottom to top in a matter of only 2 years and this rate was kept up for about six years. The good times came back to the countryside with a vengeance. The spin off into the other trades helped the rural areas to prosper, from blacksmith, to motor mechanic to seedsman. Banks were more inclined to lend money as they could see a good return with the government-backed prices making farming a gilt-edged stock. Although the war ended in 1918, such was the state of the world that food was still difficult to get hold of and shipping was still several years from reaching its pre-war tonnage.

However, by 1921 things began to change, but not for the better as far as farming was concerned. The government soon realised that they could not go on supporting the price of wheat. It was getting cheaper on the world market as the New World was exporting again. So the support rug was pulled out from under farming's feet in 1921 and within a very short time the capital value of farms and its stock- in-trade was halved.

For the men back from the war and starting a career in agriculture it was a disaster. They had bought carthorses at £100 each only to find they were worth only £50 after only a very short while. The same applied to land and, although many survived, many failed and fell into bankruptcy – again, a repeat to the aftermath of the defeat of Napoleon. As was the case then, those that had thought the farming life easy had overspent on the luxuries and paid the price.

The food producers of the world were also in financial trouble. The Empire which had helped win the war was now in trouble, through lack of money. Britain which was investing a great deal of money in the undeveloped areas of the world, now found that her investments were worth very little and quite naturally stopped paying out and started saving instead. The Empire and Argentina then started to sell their products for next to nothing just to survive.

The Government could then not resist going back to 'the free-trade' principle of the 1840s by buying the cheapest food available. If British farming could not compete then it was too bad. It was felt that it was better to sell manufactured goods and buy in cheap food, than to go down the subsidy pathway. The howl from the countryside soon reached Parliament but for a while it was either not listened to or not heard. The free traders held them off but gradually the landowners began to gain a little in the way of concessions.

The big estates, the Universities and the Church which were all massive owners of land, all needed good rents to make their system work and they had enough muscle politically to be heard. Mind you, other industries were also trying to compete for money from the government to help the basic wealth producers such as coal mining, ship building and cotton spinning to carry on, albeit uneconomically. Agriculture, oddly enough, came out of the debate far better than some of these others. Possibly because the nation felt that the rural areas were the bedrock of a sound nation. Men of the soil, the honest peasant must be helped to keep going.

So by the early 1930s wheat growing got help in the form of subsidies and the growing of sugar beet was also encouraged, both gaining some £5 million of taxpayers' money. All this helped the farming disaster areas of the Eastern counties, Norfolk in particular. There was no doubt that things had come to a very poor state in this part of the country. Large parts were no longer being farmed or if they were, tenants could not afford to pay rent. Landlords could not keep up with the repairs and their only hope was that the tenant would at least keep the weeds down so they could hang on until things improved.

There was little choice as they could not sell even if they wanted to because no one wanted to buy. This pattern ran well into the middle of England and into the arable parts of the south but began to peter out as one went to the West where grass was king, not wheat. However, even here what ploughed fields there were in the aftermath of the war, fell back to weeds rather than being put back to grass.

As happened in all economic depressions young people with muscle came along who felt that they could see a way of making a living. The old and the traditionalist began to fade away just as the upturn was coming. The young felt that land and rents could not go lower and now was the time to jump in. If you did not have any money at least if you had simple tastes and with plenty of muscle it could still be a good life.

CHAPTER 13. A REFLECTION ON THE FARMING CHANGES IN THE 1930s.

Perhaps it is time to reflect on the 1930s era before we can understand the coming war years.
If one accepts that the 1930s was the very bottom of the great depression, from then on but oh so slowly, things for the man in the street began to improve.

The heavy industries were not at the forefront of this improvement, so unemployment there was still very great. However new homes were being built and the new-fangled electrical factories were finding great demand for cookers, toasters, heaters, irons and so on. Car sales were also increasing. Farming was shedding labour at almost 10,000 men a year but agriculture was not being altogether left out for, in the richer south, a demand was beginning to come along for fresh foods such as fresh eggs and milk together with bacon.

Now the question was could the farmers take advantage of the new opportunities coming along? In a word they could not. They lacked any marketing sense, selling from hand to mouth and with no attempt at getting a standard product to the housewife. To fill the void in good marketing, the Danes came along and, like the Viking invasion of a thousand years before, it took the British farmers breath away. The government of the day tried to help British agriculture in various ways but it was rather too piecemeal to have any effect. The cry "Buy British", with posters displayed everywhere to encourage this, failed. The free trade ethic was too entrenched in the housewife's mind and purse for her to care about the fate of the farmers and she was quite right. British farming had to pull its finger out to compete but it had no firm base on which to begin, unlike the Danes and farmers from other countries who had been encouraged to form Co-ops and take on board the discipline needed for them to work.

At this point I must mention another new factor that was coming to the fore on the farming front – the National Farmers' Union or NFU. Tentatively started before the First World War by one or two brains, it picked up members in leaps and bounds in the depression years when farmers needed help from the powers that be. The men at the top of the organisation were able to debate well and many had contacts in the government via the landed gentry. They lobbied the government for an answer to the Danish invasion, and other threats, and they did this by proposing the establishment of Marketing Boards which would control the farm produce from the farm gate to the market.

Parliament was not prepared to force this on the country. However it did lay down that before the Boards came into being there must be a majority vote for them from the farming community. One would have thought that this idea would have been voted for overwhelmingly by farmers, but it was not so. It took a great deal of fierce argument at many meetings across the country before the narrow vote in favour which enabled the boards to come into being.

This exercise showed only too clearly that the British farmer was not prepared to give up his individual marketing ways in the interests of forwarding the industry in getting a major share of market. This 'head in the sand' approach has hampered British farming right up until the present time. It is probably worth pointing out at this stage in the farming story that the continentals had always had the backing of their governments. Although maybe the way the land system worked in Europe and the long distances involved on land rather than sea, together with the lack of an empire had much to do with the stable way of life that producers seemed to enjoy.

In Europe, agriculture was fundamentally peasant farming. In France, there were small farms strip farming, as in Middle age England, until Napoleon's time. Ownership of the land after the French Revolution was family based and amalgamations were certainly few and far between. The country did not go through the trauma of the Enclosures Act. It was only the coming of Napoleon, who could see that strip farming was wasteful, that brought the legislation to consolidate the strips into more manageable units, making the system more efficient.

Added to this was the fact that the British blockaded the continent for almost 20 years during the Napoleonic war in their efforts to defeat the French. It was an economic stranglehold so the Europeans looked to become self-sufficient. The blockade played into the hands of the British farmer. Prior to the blockade, when wheat harvests were poor in England, the wheat would be imported from France which tended to keep the prices stable. With the coming of war this practice stopped and the price of wheat rose.

Up until the Napoleonic wars, France had imported sugar from its West Indian islands as did Britain. When this supply route was cut off they had to put their minds to satisfying the sweet tooth of the nation another way. They did this by breeding up a beet plant which could take the place of sugar cane and sugar production is now one of the biggest industries in that country to this day. That, together with the grape, kept the spirit of French Agriculture in good shape year in year out.

As a result of the strip farming amalgamations, the French peasant was only paying rent to his family, and was too tough a customer to be beaten by the harsh economic times. Under stress, belts could be tightened even further. The British had developed bigger farms, employed labour and paid rent to a landowner who wanted his money come what may, which meant that belt tightening was not so easy. Canada and the Americas had been lost to the French (look up the story of the Canadian wars and the loss of land in the south of the USA) and the great explosion of food they produced did not have as great an impact on the French as it did on the British.

In Britain during the 1930s, the countryside changed dramatically from every point of view. It was an environmentalist's dream although not many people at that time were concerned with that aspect of life. The once well farmed downland, well stocked with sheep and into a four course rotation, went back to scrub and brush. It was used for grazing on the cheap - one took what grew and expended nothing on it. Fences and hedges were left to decay. The difficult land such as heavy clay soil which was cold and wet was neglected and returned to rushes. Ditches and land drains, put in at great expense in the 1860s, became blocked, in its way a National Disaster.

Around the time of the Boer war the physical health of the army volunteers was very much under par. It was no surprise when you consider the background from which the recruits came. Poor housing, a population explosion, together with the lack of a decent diet as a result of the shift from country to town due to the Industrial Revolution. At this time a focus in the health of the nation began and the coming of the Town and District councils in the 1900s and the increase in their power in looking after the interest of its local people also played a part. At the start of the First World War the poor state of the recruits once again, really pushed the nation into taking its health seriously. It was a very slow process, which began with the collection of country wide statistics on health. The worrying figures that began to come out of these statistics warranted investigation. An interesting fact emerged after the introduction of food rationing during World War One, which was that conscripts in 1918 were far fitter and heavier than volunteers at the beginning of the war. However these findings only began to have a true impact on farming from 1930 onwards. Together with the introduction of the Milk Marketing Board (MMB) this was the start of the big expansion in the Dairy industry.

A pilot experiment by a County Council in the industrial heartland of England was the final catalyst that really set the pace of change in Dairying. All school children were given a 1/3rd of a pint of free milk every day. Their weight and any deformities were recorded from the beginning to the end. The result after a period of time proved beyond doubt that this milk input was of tremendous value. The scourge of rickets and other bone abnormalities receded. The children's weight for age improved without question and so began the School Milk scheme. This continued until the 1990s when the nation felt that the general diet of children had improved sufficiently and it could no longer afford the cost.

Dairy farmers who had a farm with a good water supply and good transport links to town were putting on more cows to satisfy the demand for fresh milk during this decade. This trend had been going on since the disasters of the 1880s but it was a cut-throat business to say the least. Perhaps it would help if the reader could understand the time it takes to get a dairy farm into operation and what it takes to keep it there.

If one takes into account the gestation period of a cow' it takes three years or more to get a dairy herd to the point where it is ready to give milk. One must also remember that only 50% of the calves are female! The cows have to be milked twice a day. The milk must be cooled to stop the growth of bacteria which cause it to sour. This was not easy in the age before cheap refrigeration. A buyer in the nearest big town then had to be found who would take the milk week in week out. This buying agreement was usually done on an annual contract. The snag was that if in the second year a farmer could not agree a price or someone undercut him, then there was nowhere to sell the milk which was coming from the cow twice a day no matter what.

The dairy farmers were too far split to organise their own marketing and consequently they were often under cutting each other on price. Many tried to convert their milk into butter or cheese to get round the problem but could not beat the producers from the Empire especially the New Zealanders, on price.

An additional problem arose which was how to maximise winter milk production. Feed costs for the oil cakes and other goodies needed are much higher than getting milk from grass. The town buyers understood that production costs were higher in winter and paid more per gallon. When spring came and the milk flowed faster (known as the spring flush) they tried to cut the farmer's price as low as they could, together with a restriction on the amount they were prepared to take. So any surplus was likely to be poured down the drain and wasted.
Clearly the free market was not working and the dairy farmers were in a very difficult financial position, many going bankrupt at a steady rate.

It was these issues that led to the introduction of the MMB in 1933. Members of the new board came from farmers, retailers, the public and government employees, all tied into the Marketing Act. This board had the power to market every gallon of milk produced, operate a pool price for farmers, however distant they were from the market, and to sell to retailers. The MMB could also charge a levy on each gallon produced which created a capital fund that could be used to build plants to make cheese or butter from any surpluses. What a revolution had come about in the marketing of this very fragile commodity!

There was much wailing and gnashing of teeth from the wholesalers, as one would expect, but gradually the benefits could be seen by all and at last the country had set itself on course to having an assured home grown food supply.

Meanwhile the continentals continued down the Co-operative route, helped by their governments and a land bank which provided loans at cheaper rates than standard.

At this time in Britain, during the 1930s, other marketing boards were created for pigs, eggs, potatoes and hops. However only the Hop Board made any progress. The others failed as they could not control the whole range of the operation. It was the Milk Marketing Board that at last put the backbone into farming – a guaranteed cheque at the end of each month was a wonderful tonic! This gave confidence to the industry and more and more folk began to go into dairying increasingly backed by the banks.

The product was being advertised over the whole of the country and the consumption of fresh milk rose year on year. Now farming cows was good business so they exploited this.

Although British farming was becoming more profitable once again, it was only at survival rates i.e. there were less numbers going under. The introduction of a subsidy on beef and wheat and sugar beet growing helped the arable areas just to keep going. If the farm had a water supply then the farmers could start a dairy herd but water in the eastern counties was in short supply and there was no money with which to put in piped water.

Another great problem was the damage that rabbits were doing to farmland. Nothing could be grown in the way of an arable crop unless the field was surrounded with wire netting. This kept the pest out to a certain extent which gave the crop a chance to come to harvest. The downlands were awash with rabbits and many tenants just kept themselves solvent by selling these rabbits which were in enough numbers to cover the rents.

On the chalk downlands there were no buildings that could be used for cows. They could only be kept on the meadows near the rivers in buildings that cost a great deal to put up. Very few landlords would find the money to put up more sheds and this made it very difficult for anyone to expand their dairy herd.

Grass was very little used as a crop for cows. Hay was made as a 'fill-belly' for the winter and what grasses the spring brought was a bonus. The foods that really made the milk flow were the imported cereals and the by-products of the seed crushing industry. They were cheap and good and the agricultural merchants were keen to sell them by the boat load. At this time in the 1930s great companies started up to formulate rations for cows so that the farmer had no need to mix his own feeds. In fact fertility from around the world was being imported into England, to put through the cows and other livestock and the rich muck was then put back on the land.

This improved fertility was not exploited at the time in question – maybe the hay crops began to get better but cereals did not come into the system as they could be imported so very cheaply what was the point?

Land prices had reached a very low point especially for those areas of chalkland that none could find a way to farm. £5-£10 an acre was about the price or alternatively a tenant could have them for free provided he kept the weeds down.

At this point the new keen young men began to come onto the scene and a chap called Hosier was the first to exploit the situation. He invented a movable milking bail that could be constructed very cheaply and put out upon these downlands. In the early stages they were used for hand milking the cows, but before long the milking machine was installed driven by a small petrol engine. The water problem was overcome by the construction of windmills on top of the farm to pump water from deep down in the chalk. As this soil is so very free draining the cows could stay out all winter and they kept very clean and healthy by always being out in the fresh air not chained up with their herd mates breathing each other's air. This point was a great bonus for the spread of tuberculosis was often very rapid in the byre housed cows.

The other bonus of course was the small capital cost to set up the milking unit. One man could milk more cows than his brother in the lowlands. He had no manure to shift each day – just the bail to move which he did with a horse. The horse also took the milk churns back to the road for collection by the milk lorry. At last the downs were getting back that fertility they so badly needed, and had been deprived of, since the sheep flocks had to leave the scene way back in the 1890s.

This man also found a way to handle poultry flocks which, up to this time, had not really progressed in anyway other than the breeding of better birds. Hosier built cheap fold units each housing about 20 hens which were then safe from the fox at last! These units were moved every day onto fresh grass thus cutting out the worm problems that were the scourge of the large static hen houses. Again imported foods were being passed by the hen onto the soil.

These new systems very soon caught on in a big way and the downs were being covered with milking bales and fold units. Cows to use them were being bought in great numbers cheaply from Ireland and were hardy Shorthorn stocks which was just what was needed on the windy downs.

In the meantime, on the other side of England in Essex, another farmer was busy with Dairy cows but in a rather different way. He had been to see cattle in Holland in the Friesland areas and had come away impressed by the great milking potential of these cows. He was in a way following in the footsteps of Turnip Townsend some 200 years before who had also had a look at the farming of Holland and had come back with Turnips and Ryegrass.

This time what was wanted was a cow that would give a great deal of milk and would breed calves which had inherited this gene. The Dutch had for some years been recording the milk yield of their herds and could now reap the benefits of selective breeding. These animals could use the imported feeds to great effect by giving some 1000 gallons of milk whereas the poor old Shorthorn struggled to achieve 500 gallons. The only snag was the very low butterfat in the milk of the Dutch cattle but as the idea was to create big milk rounds on the fringes of London this was not so important.

Yet another farmer was also seizing the chance to make money this time in the arable field. He could see the potential of introducing the prairies mechanisation onto the big open fields together with the use of artificial manures, which were starting to come onto the market, to cheapen the production costs of cereals. He introduced the use of sulphate of ammonia, a by-product of the gas industry, which gave him the nitrogen that this crop needed.
The big snag was the rather damp climate of this country and his problem was how he could get the grain dry so that it would keep. He overcame this by inventing a corn drier that was heated by coke, a cheap fuel which was another by-product of the gas industry.

At this stage all these happenings were on a small scale compared to farming as a whole but the new ideas were to become an integral part of farming history and the geographical split in types of agricultural production. As you travel west and north in the country the climate becomes more humid and the rainfall increases. This climate change does no favours to the production of wheat and barley but is liked by the grass species. Thus farming turns on the ability of the farmer to turn his grass into profit with livestock.

This leads me onto Robert Elliot who was a prophet before his time and demonstrated the use of grass to improve the fertility of the soil. He was a landlord rather in the mould of Coke, for his vision of how to use grass to maintain and improve the fertility of the soil. The estate he owned was in Roxburghshire and he took the worst farm in hand to carry out his idea. This all happened before 1912, well before the low point in farming that followed the opening up of the plains of the New World. He showed that by using a six year rotation the soil fertility could be greatly improved. This rotation would be as follows. In the first year, he would sow grass with clover, the mineral rich plant, and this would remain for three years (a three year ley). This was then ploughed up and followed for 2 years by a cereal crop. In the sixth year he would sow roots. Using this rotation, it was possible to get good yields without the use of very much artificial manure. Of course it was the stock grazing the three year ley that was the key to the whole system for it was the manure and urine that stoked up the soil fertility for use later.

Elliott felt that this was the only way that British farming could compete with cheap imported cereals. If the soil was improved, as the manufacturing base of the country was declining leading to reduced finance, the nation would be able to feed itself rather than have to spend any money abroad. It was great stuff to tell the farmers and he proved that it worked. His idea began to filter through to the Board of Agriculture who gave him the great sum of £50 a year to help with his experiments and records.

Then came Stapleton, another man of vision. He took the grass story by the scruff of the neck and revolutionised its productivity. Up to that point hardly any grass seed varieties other than Italian Rye grass were available to the farmer. If he needed to return from arable to grass for whatever reason, he had no good grass seeds to achieve this. He could only let the land fall back to weed grasses or sweep out his hayloft and scatter the seed on the plough.

Stapleton and others, who were doing nutritional work with grass, soon proved that it was the leaf of the plant that was the key to better milk production and weight gain in sheep and cattle. But the leafy plants would have to be found and then bred up to produce a stock of seed that could be relied upon. Stapleton discovered that there was a huge range of grass species to look at each one having a characteristic that could well fit into a seed mixture and give the best feed value for a greater length of the growing season. Unlike cereals, no one knew how to manage grass so that it could be kept at the leafy stage as long as possible. Grass only wants to do one thing in June and that is to go to seed. When it does this the feed value goes down like a punctured balloon. No one had thought of the best way to manure grass unlike the arable crops that Laws had looked at back in 1840. No one had done any real research work on the best way to cultivate the crop. So much work needed to be done - much of it in the dark so to speak, but not quite.

Just at the wrong time along came the Second World War. All hands to the plough for grain to feed the nation. No one was looking to the future except Stapleton.

CHAPTER 14. THE WAR YEARS OF 1939 TO 1945

The state of play in Britain in 1939 as I have tried to show in the last chapter meant that new ideas were coming into place and mechanisation was on the point of rapid expansion. The need to tackle the health of the nation's dairy herds was being discussed. In fact the farmers of Britain were on the brink of competing with the farmers of the world and in a position to chase them on price, a position they had not reached since the Empire had wound itself up to earn money from the old country.

September 3rd 1939, the day war was declared, put paid to all of this. It was a case of 'up corn and down horn' again! Back to the regime, as in WW1, of filling the bellies of the nation with a bland but wholesome diet. No one in that fatal year had any idea how long the war would last. After the fall of Poland, which was achieved in only a matter of weeks, a sort of stalemate took place called a phoney war at the time.

However, by May 1940 things began to get very hot. Unbelievably France fell and England only just hung on thanks to winning the Battle of Britain in September of that year. No-one except the Commonwealth came to our aid in any belligerent fashion. Then the big battle of the Atlantic began in the winter of 1940/41, after the Germans found no peace feelers coming from the British Government. Ships were being sunk in ever increasing numbers.

Rationing, which had begun in 1939, started to become more severe and the time came to really speed the plough. Beef and Mutton together with Pig meat production had to be cut way down. Cereal and vegetable production and as much milk as could be obtained from grass and fodder crops was the order of the day.

The country was in no doubt now that it would be a long haul to victory. Food was the crucial component to enable the island to hang on until something or someone came to the rescue. By June 1941 Russia was invaded by the Germans and they also decided that they could crack the British Isles by stepping up the destruction of shipping. Generally speaking the country needed about a million tonnes of supplies each week. Almost all was coming from across the wide Atlantic ocean. The navy was finding it almost impossible to police the vast pond and the U boats found they could sink ships right on the doorstep of the USA. Britain was in extreme danger. It was only thanks to the President of the USA, who began to see the danger to his country if the Nazis got world domination, that complete disaster was averted.

He, the President, was treading on very thin ice in trying to help Britain. Not many in the States wanted war but he did get away with attacking the Germans when they sank ships near the coast of his country. That helped, but the enemy soon brought out new submarines and began to attack the convoys in large packs and they could do great damage in only a night or two in the mid-Atlantic.

The larders of Britain began to run down week by week, for it takes time for things to grow and ploughs to plough, even though a start was made in 1939. The Nation was in a frightening position - one might say desperate.

Now let's try and understand what was being attempted on the fields of England. Unlike the position in the First World War, plans had been set afoot well in advance of the declaration of war. Grants of some £2 per acre were made in 1937 to get the plough turning the grass sod. At that time it was still the farmer' choice as to whether he took to the plough.

The farmers of England were mostly tenants and tied to landlords and tenancy leases. Under the terms of their tenancy agreement, most tenants could not plough their fields. This was the Landlord's way of trying to make sure that there was not a return to the old ways of the 1870s when arable farmers failed to pay their rents due to the low price of grain.

The government of the day soon swept this problem away using The Defence of the Realm Act. Orders to plough came pouring out of Agricultural County Offices. To start with the sudden imposition of orders onto farmers, who had managed to overcome the depression of the thirties by not doing any more than was needed to survive, did not go down well. To be given orders by their fellows was a thorn which pricked pretty badly.

Arguments, appeals and refusals to cooperate were common-place. The phoney war did not help the situation. What was wanted was a sign that there was a war going on. It came with a bang with the fall of France, the battle of Britain and the start of the battle of the Atlantic. The rationing of food together with that of fuel began to bite deeper. Mix all this in with the fear of invasion and farmers, let alone the rest of the country, finally got the message.
Once the nation was at war things changed.. To facilitate the new situation, each county had been given an Agricultural Committee appointed by the government and composed of farmers and land owners. Their job was to transform the orders from the government into results from the farmers within their county.

A further sub-division was made of the county, into district committees again with farmers appointed onto them by the County committee. For example the government calculated the total area of wheat that was required and then that information was passed down to the district committees who had the job of going to each farmer within their area and telling them what they had to plough up to achieve the target set. This proved to be no easy task for all sorts of problems came to the fore such as the lack of machines to do the work of ploughing.

On the farms that had been entirely devoted to livestock, farmers realised that this meant that they would be short of grass and were afraid that they would have to sell their stock. Strict rationing of concentrated feeds that came from abroad, meant the Pig and Poultry farmers had to reduce their herds and flocks for the number one priority crop, wheat. The dairy cow suffered less than other stock as the nation needed milk to make up for the loss of other proteins especially for children and nursing mothers. The lessons of the thirties had been well and truly learnt.

However the cow was not entirely unaffected. The concentrated feeds on which she had depended in such large amounts before the war were severely rationed which meant her milk yield was bound to reduce. With luck, she might manage to produce enough liquid milk for the nation, but any cheese or butter production was out.

Whether the government of the 1930s realised that the war with Germany was inevitable I do not know, but around 1937 more of the intellectual elements within farming were beginning to debate how Agriculture might respond. One of these men was Stapleton who, in the 1920s, had become the first Director of a grassland research station set up at Aberystwyth in Wales almost certainly using Elliot's ideas.

Stapleton was getting a name for himself on how to go about farming grass. A few farmers were having a go with his ley farming recommendations and getting results. Now he could see that, should war come again, farming grass would be put on the government's back burner. But he could see a place for grass in rather an interesting way. The main thrust of his argument was that if grass could be made to become far more productive it would release more ground to the growing of human food crops. A ley system (putting grass in a rotation) could also help to improve their yields. Most importantly the poor uplands and mountain areas had a part to play in the war scene. If these pastures could be improved by manuring and renovating with new grasses that he had bred up at his research station, then it was too good a weapon to be left out of the war armoury.

By 1937 the government made a move by giving £2 an acre to farmers who ploughed up any old or worn out grass, put on subsidised lime or basic slag to put right the fertility.

This move was too late to get the show on the road for by 1939 we were at war with not much grass in a good productive state and the import of animal feedingstuffs declining rapidly. Farming was still in a rundown condition, it was not in quite such a mess as it was before 1930 as I have tried to indicate, but it was not ready for the war either. Certainly it might have been if more steps had been taken by 1935.

However very important lessons had been learnt from the First World War and the organisational capability was in place to carry out the resurrection of the farmlands. It was brought into force right from the word go and to a certain degree some stockpiling of food, fertilisers and tractors had already taken place. Perhaps one of the most urgent deficiencies was that of farming academics to put the new ideas into place. So folk were culled from Universities and farm schools, retired academics and the best from the Agricultural Merchants.

Conscription was brought in in 1938 and farming was once again declared a reserved occupation. With all this put into place it still was felt certain that by 1942 experts from the Empire, New Zealand in particular, would be needed to help out with the Livestock section of farming. Some had already joined the army and were asked to return to the farming front. The direction given by the War Agriculture Executive Committees (WAECs) as dictated by the Minister of Food was to grow as much wheat as could be managed, put in yet more potatoes, (a crop which needed much labour and not well–liked by many farmers) and increase the sugar beet crop. This sounds simple but in practice it was far from it.

Let us start with the land. There were many problems to deal with, the first being drainage. Vast areas had been drained in the Golden years of the mid-1880s when millions of pounds of government and private capital had been spent on the drains. In the depressed years from1880 onwards the simple task of keeping ditches cleaned out to let the tile drain run had been let go, the tiles had blocked up and the soil had become waterlogged once more.
The next problem was the acidity of large areas of land again due to lack of money available to apply chalk to the soil. Problem number 3 was the lack of sheep on the downlands which used to maintain the plant potash needed for the cereal crops. Cows had certainly helped this but not in large enough numbers to relieve the situation. Problem number 4 was the shortage of arable machinery to do the work and also the men who knew how to operate it. One could go on but I hope I have given an idea of the difficulties that were cropping up day in day out. The knowledge was abroad, on how to grow more, as had been shown by a few farmers

back in the late 1930s but it needed spreading. So the WAECs not only had the task of enforcing the orders on farmers and trying to educate them, they also had to create machinery pools which could go out to plough and sow the small farmer's land which would otherwise be left uncultivated through lack of machinery. The committee members not only had to manage their own farms but also those that were under their control. There is little doubt that these men worked all the hours that God made to get the production up to the mark.

So it can be seen that farming generally was not geared up to the arable system and yet that was what the government wanted. So many farmers were in need of retraining in the skills of arable farming including the types of fertilisers to use and the best varieties of cereals to put in. Some also needed help in the use of the plough and the art of building ricks that would keep out the weather. The District Officers, who had been appointed to help, found it hard work to cope with the demands asked of them. As often happens in this country a self-help system began to be developed by the more progressive farmers. It came in the form of Agricultural Clubs often called Growmore Clubs. At the meetings held, lectures would be given on the latest techniques in crop production and also demonstrations on how to set a plough and how to plough. Anyone who had a good idea on the farming front was more than happy to share his knowledge, keeping nothing to himself. It was, in its way, a revival of the Agricultural clubs set up in the early 1800s.

These Growmore clubs developed at Parish level with the aim of raising the farming output, often holding local competitions such as ploughing matches.

There was little doubt that it worked and helped the government achieve its aims. Patriotism played its part and the profits of course, but the feeling that at last farming, British farming was wanted and could not be ignored, gave hope after such a long time in the desert of the Depression. After a while the farmers began to see benefit in all of this activity and began to forget the frugal methods of the depression and spent money on their farms. There is nothing like a willing horse to get the work done.

1942 – 1943 was one of the blackest periods of the war. Nothing was going right except that at last the Americans had joined the fray, thanks to the Japanese. Fuel and food supplies were still on the knife edge of a disaster and the battle of the Atlantic raged on with great losses of both men and ships.

This meant that supplies for farming needs fell behind those of the Army just at the point when they were needed on the farm. The numbers of tractors had been on the increase before the War but what was lacking were the implements to pull behind them. Much of the imported farm machinery from America was being sunk and only a little trickle was getting into England.
So then for farmers, it was a case of having to make do and mend. Here the blacksmiths came to the fore by learning how to use welding equipment. They set about converting horse-drawn machinery to be pulled by the tractors. Shafts were taken off and drawbars put in place. However they could not keep up with the work that the farmers wanted done so progress was slow.

Harvesting still had to go on and this was helped by the setting up of holiday camps so that folk from the towns could help with the cereal and potato harvests which were vital for keeping the bellies of the nation full.

Rationing in the First World War (WW1), was brought in quite late in the proceedings, and seemed somewhat defeated by the fact that if you had money you could beat the system. This time rationing began early and the system was so arranged that everyone had a fair chance around the National table. A lot had been learnt about healthy eating and children in particular were given most of the protein and vitamins that were available.

The rosehip was another harvest that was entirely provided by Nature and went a long way in replacing fresh fruit. Many thousands of tonnes of rosehips were picked from the hedges and byways, sent to collection depots provided by the Ministry of Food, and then crushed to extract the syrup.

CHAPTER 15. THE CHANGING EMPHASIS TO LIVESTOCK

If there was one thing more than any other that helped farming feed the population it was the inherent fertility of the soils. Not that High farming, as demonstrated by the grand old men in the 1840s, was the reason. The change from that system in 1880 to importing the foods of the world in the form of Oil seeds and imported cereals was the main reason. These foods were fed to all species of stock which then produced that for which they had no need, depositing year after year this excess fertility onto the ground. Perhaps this is rather a generalist view but there was no doubt that this was happening. Farmers were not taking anything like as much out of the soil as they were putting in. It was cheaper, it seemed, to buy in foodstuffs than to grow the good grass leys that Elliott and Stapleton had demonstrated could be grown.

War suddenly stopped this and the ploughs began to turn up the old pasturelands and convert them into cereals. Thus the imported fertility together with what little artificial manures could be obtained, saved the day. However the pioneers of the new farming techniques in the pre-war years began to have the ear of the government, pointing out that farming needed to replace this fertility. They knew how it could be done using the cow and well managed grass.
If, they said, all of this was combined with improving the genetic makeup of the cow then milk output would be increased. The cereal output could be maintained and the health of farming and the nation could be improved.

This change of thinking began around 1944. What triggered this change in policy, when it seemed that the war was being won and that the pre-war system might come back, is not easy to explain. It was probably due to lack of dollars, for Britain had cashed in most of its assets to fight the war. The major factor to be considered was the cost of war, a war that engulfed the whole world which meant that food would be short and prices high. There was very little available to buy compared to the pre-war days, and anyway there was very little if any cash in the kitty to spend. No doubt as the Battle of the Atlantic was being won, grain could come in from the prairies once again but Europe was devastated and the war against Japan was yet to be tackled.

In view of this great encouragement was now being given to ley farming, the use of clovers in the leys and the good management of the sward. The main problem to overcome was the production of good grass seed to sow these leys with. Now the Welsh grass station in Aberystwyth came into its own. The ideas that had been bubbling, or perhaps we should say simmering, before the war came into the forefront of government thinking. Stapleton was in great demand, to explain to the farmers his philosophy of the continued recycling of grass through the animal leading to the build-up of fertility together with that other soil essential humus. Increasing the production of quality grass meant more animals could be carried per acre than was the case with the permanent pastures. The winter fodder resulting from these leys was far better than could be obtained from poor unimproved pastures and, together with the limited amount of imported feed he was allowed, the Dairy farmer was a better position to produce more winter milk.

Slowly farmers began to heed the new ways, but the tool they had to work with in the form of a cow was in need of an overhaul. The genetic makeup of the cows used for milk production and their health was not robust. It was to be tackled.

Breed Societies that had sprung up from way back in the 1880's found much more fertile ground now than they had done during the depression years and could see that they had to meet the demand for improvement in the cow's milk production.

There was money now in pedigree animals especially cattle, now that farmers began to feel profits could be gained. The Milk Marketing Board began to run the National Milk Record Service that could show the true yields of cows rather than the yields recorded by their owners, the accuracy of which could be doubtful. Thus at this time, the National Dairy Herd was beginning its modernisation, or at least the bones on which progress could be built. One thing was lacking, and that was the way to improve the cattle faster. At that time a bull could only be used in one herd at a time, and then could only cover some 40 cows so progress could not be achieved on such a restricted basis even with the use of good bulls.

In 1944 the breakthrough came from the Cambridge Animal Genetic Department, in the form of Artificial Insemination (A.I.). Thus one good bull was able to give enough semen to inseminate some several hundred cows each year. Again farmers very soon found a way to exploit this new tool to their advantage. They formed co-operatives that found the finance to establish centres where bulls could be kept and from which inseminators could travel the district and service cows in large or small herds. The men doing this job were often known as the 'Bulls in the Bowler Hats'.

A.I. also had another role to play in the fight against sexually transmitted diseases in cattle. Areas in which small herds were predominant, such as the small hill farms and those in the New Forest, often shared bulls and found this disease problem was causing great economic difficulties. At the very beginning of this A.I. invention the Ministry of Agriculture made sure that the veterinary service laid down very strict ground rules in its operation. Not only from the disease point of view but from the conformation angle, as great damage could have been done if the bulls used had deformities, however slight.

Although some farmers felt that A.I. was cheating nature and would have nothing to do with this 'new-fangled' idea the system moved on at a great pace as most farmers realised the great potential they had at hand. The upsurge in the demand for pedigree bulls from milk recorded dams to go to the A.I. Centres made many breeders rich men. The groundwork of the depression years carried out by men of vision was at last coming to its peak. The cow was at last Queen at this time and the work on sheep, pigs and poultry was very far behind. Grain production was not good enough to provide a surplus that could be channelled to feed them. It would come, but not until the 1950's, some ten years away.

CHAPTER 16. THE ARABLE STORY

At this stage in the war, which was still raging away all over the world, farmers were lapping up all they could in the way of new agricultural knowledge. They knew that food was desperately needed and they, at last, had a chance to revitalise British farming and put a healthy glow on its cheeks!

It might at this stage in this brief history be of interest to detail some of the new ideas that were coming forward in the arable sector of the war front. Livestock as we have seen had its new clothes made for it not long before 1939 but was unable to wear many of them until towards the end of the war. Arable men had to wait some ten years for the new research to bring to light a better road way for them.

No doubt mechanisation in harvesting the cereals started before the war by one man showing what could be done with a combine harvester on the cheap downlands. But the tractors pulling the ploughs were only pulling adapted horse ploughs, and the big multi-furrow ploughs of the steam era had long gone in the depressed arable period.

However the fundamental method of growing cereal crops and roots had not changed since Jethro Tull came up with the drill. His objective was to enable the cultivator to hoe out the weeds easily so that the crop had very little competition for light, moisture and nutrients.

The huge area of the chalk downlands, easily worked, failed until sheep and root crops were introduced, together with the feeding of oil cakes. The treading of the sheep and the release of plant foods readily available to the plants, via their alimentary canals made certain the barley crop yielded well. But by 1935 the hurdle flocks had become uneconomic and the milking bails could not seem to produce the same results.

War time came to the rescue and started the revolution in downland farming that has lasted to the present day. It hinged on making the best use of scarce artificial manures, and was an adaption of the corn drill to sow fertilisers down a tube and thus close to the seed.

Due to their alkalinity, the chalk downlands suffered from a shortage of potash in solution which meaning that the potash was not available to the plants. Putting the potash close to the seed overcame this problem and the crops of wheat and barley soon showed what they could do. This was called combine drilling.

At about the same time, the Imperial Chemical Industry (I.C.I.) which was the biggest chemical firm in the country, possibly the world, was well entrenched in producing artificial manures and then began to work on weed control in the cereal crop.

Now the weeds that caused most havoc, certainly on the downs and gravel soils were charlock and poppies. Their rate of growth was so fast that within a few weeks of the cereals germinating they would smother the lot, cutting yield by 50%. The early attempt at chemical control was by the use of sulphuric acid, mixed with water and sprayed on the crop. It worked because the acid would attack or burn the charlock but because the cereal plant was slightly waxier would do little harm to them. It was quite a job handling this acid, to say the least, especially without proper clothes. The sprayers of the day were not very efficient or accurate in their application. No tractor sprayers were about so the carter and his horse had to do the job. At this time it was not easy to find water up on the downlands, and as some 100 gallons to the acre had to be applied, it can be seen that the task was beyond the scope of most farmers.

I.C.I. came up with a product derived from their work on plant growth hormones. Little water was needed to get a good spread of the product on the crop. The material made the charlock and poppies grow far too fast and so they killed themselves and the corn was in no way affected. Needless to say the yields went up and harvesting weed free crops was a great bonus.

The combine harvester was another big development but the climate in this country was a factor that made its adoption slow to be taken up. The combine could harvest the grain when the grain moisture could be well in excess of 18%. If grain is kept in large heaps or in bulk then the moisture must drop to 15% or below to prevent it going mouldy. However in a hessian sack it can be stored at a moisture level of 18%. In a rick made of sheaves which were harvested in the old fashioned way, the moisture of the grain may be higher than 18% when first harvested but the wind blowing through the rick dried it so that by January it was in a fit state to thresh and store in '4 bushelweight' sacks.

To overcome the problem efforts were made to artificially dry the corn with heat from coal or coke. These driers were not very efficient or cheap to make. At the end of the war a surge of effort, using oil began to make them far more efficient and acceptable, and as the price of grain was good so more farmers began to use them. The use of Combine harvesters increased as a result.

It was the young men who had come into farming around about 1934/5 who led the way forward. Their older brethren still felt bitter after the let down from World War 1 and when they had had to hang on during the depression. The young chaps could see how mechanisation used on the land that was cheap to rent and buy would work wonders for the bank balance. The war held them up for a while but when the strings of government began to be relaxed they forged ahead. As happened back in the 1880's it was the United States that led the way in mechanising the harvest. First with the Massey Harris binder, doing away with the scythe, and then in the 1940's the combine harvester which did away with the threshing machine.

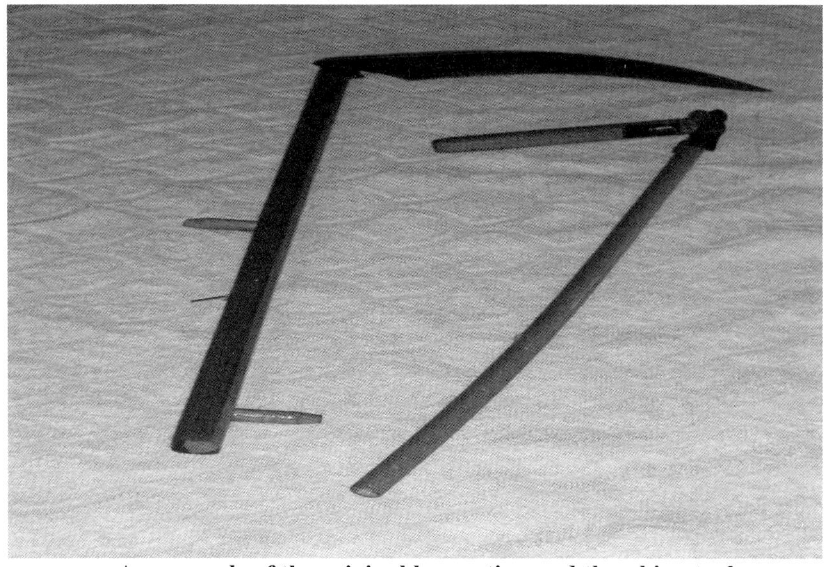

An example of the original harvesting and threshing tools.

A Horse Binder for making sheaves – imagine cutting this field with a scythe and making the sheaves by hand!

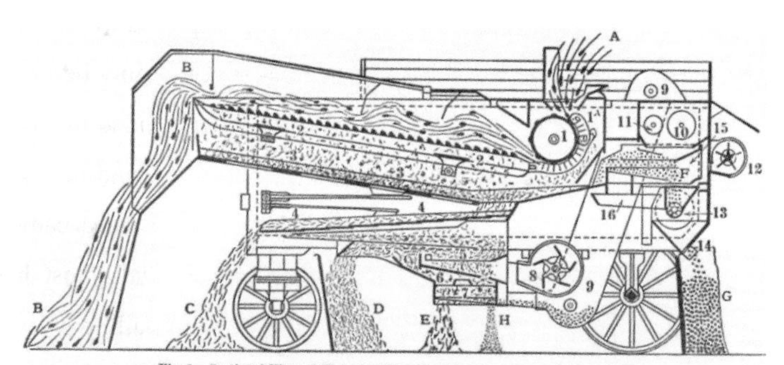

Fig. 1.—Sectional View of Finishing Threshing Machine (Brown & May, Ltd.)

A, Unthreshed Corn. B, Straw. C, Cavings. D, Chaff. E, Chobs. F, Corn. G, Finished Grain. H, Dust. 1, Drum. 1A, Concave. 2, Shakers. 3, Upper Shoe. 4, Caving Riddle. 5, Chaff Riddle. 6, Chob Riddle. 7, Seed Sieve. 8, First Blower. 9, Corn Elevator. 10, Awner and Polisher. 11, Conveyor Worm. 12, Back Blower. 13, Separating Screen. 14, Corn Spout. 15, Second Dresser. 16, Third Dresser.

Threshing machine – overtaken by the Combine Harvester

A Combine Harvester – superseding the Binder and sheaf era

One thing that was invented within the British Isles that provided a dramatic break through in the output of the humble tractor was the integration of hydraulics into the machine.

Up until 1946, when Harry Ferguson produced his tractor combination, virtually all tractors were pulling adapted horse implements or big imported implements from U.S.A. and Canada. The snag was that these machines could only be used on the bigger open farms, not on the smaller units with fields surrounded by hedgerows. Now Ferguson overcame this, his tractors sold like hot cakes, and his patent was sought by all the other tractor producers. The day this new idea came on the market was almost certainly the day the poor old horse finally lost his slight foothold within British Farming. One of these little tractors pulling a two-furrow plough could turn over in one day some 10 acres of land, compared with one acre by two horses. The manoeuvre of these machines was such that the smallest field of the most awkward shape was no problem. Ferguson, in fact got the notion from his own countryside of Northern Ireland, with its small fields and slopes. In addition, the weight transference that could be achieved from the plough to the back wheels of the tractor meant that much more wet and heavy land could be tackled than ever before.

With the Second World War coming to an end, the factories that had had to feed the demands of the army, navy and air force were well able to get to grips with the peace. The cry was now to forge weapons into plough shares at a great rate, not to save the British nation from defeat but to save Europe from mass starvation, and thus the road to beat Communism began.

The Government which had begun to encourage the return to pre-war animal production and cheap imports of grain back in 1945 had to change course to get still more cereals home grown, more in the British loaf. During the whole war period bread was never rationed but from 1947 to 1950 this changed as Britain was being denied American grain which was destined for Europe.

As had happened in winning the war it was the Americans who came forward to save the peace, this time with food and money. The Russians, who had been our allies during the defeat of Germany, now had such a hold on conquered lands in the East of Europe, and commanded such a large army, that they were in a position to march all the way to the channel. The temptation to put their communism on the map was going to be difficult for them to resist. The Western allies felt that the starving population of Europe must be fed and watered to prevent them being tempted by the lure of communism.

To achieve this the U.S.A. put their farming into top gear or perhaps one should say fifth gear. They lent the European nations millions upon millions of dollars with which to buy their grain, and to begin the restructuring of their nations. It was called 'Marshall Aid' after the man who was given the massive task of organising it.

Britain was included in this aid programme but she was still in debt to the Americans over the Lend Lease arrangements of the Roosevelt years. What dollars came our way were needed for things other than food, so it was a case of back to the land. All the main staple foods meat, milk, bread, sugar and eggs remained on ration now. Only the humble 'spud' was more or less freely available. Although the British nation had had to endure food rationing for some ten years, they had not starved and their regard for the people who worked the land was great. However, it only took some five years after rationing ended for things to change back to envy and contempt.

By 1947 the labour position on farms had certainly improved, and the chance to buy new and improved equipment was helped by the government giving priority to farming needs. Artificial fertilisers became easier to buy. Phosphates were taken off ration and the use of Nitrogen via Sulphate of Ammonia from the gas works and Ammonium Nitrate from the big chemical firms was encouraged, with a subsidy to reduce the price. In modern parlance it was 'all systems go' from grass to cereals to cabbages.

However, a stutter was on the cards. For a good number of the farming community could still smell the rat of 1920, when the government suddenly stopped subsidies almost overnight and were worried that it could happen again. The National Farmers Union, the organisation that had spread its wings in the 1930's, had played a major role during the war years, not only looking after farmers interests, but in helping the government get its orders over to the countryside. So now the N.F.U.

which contained powerful landowners with political know-how, felt it was the time to try to get farming on the national agenda to secure its future.

In 1945 a general election was held, the first one for some years. The Labour party won hands down with such a majority that they were able to carry out all of their policies with no opposition from the Conservatives. They, the Labour Party, had ideas for the future of farming in the U.K. and the result was the famous 1947 Agriculture Act. The N.F.U. was able to extract much from this party and have it incorporated within the Act. In essence they obtained guaranteed prices that could be negotiated each year with the objective of producing 60% of the Nation's food, together with a much more secure future for tenants in a new Agricultural Holding Act. This meant that the tenants could go forward with great confidence and even borrow on their security. Now the future could be tackled.

So the scene was set for the next decade and farmers lapped up the opportunity to really farm well at last. They got back into using rotations, breeding better cows, and started up the sheep and pig herds once again. With good profits coming along and a secure tenancy they beavered away at reclaiming more land, draining more land and covering it with lime. Water was laid on to the most remote fields allowing cattle to be spread around the farm. Most subsidies covered up to 50% of the cost of this work. There was no doubt that it was a boom time for farming, there was plenty of labour to hand and wages comparatively low.

By 1953, meat and bread came off ration and there was no fear of a threat from imported food. Because the war was totally world-wide, food was short, and only North America was able to export any amount of grain, and that was by giving credit in huge amount to those in need. Canada and New Zealand filled up the

30% not covered by U.K. farmers and they were able to make a good living, compared with the aftermath of the First World War.

The War Agricultural Committees, set up over ten years before, were wound down. The Minister of Agriculture kept area liaison committees, recruited from the more progressive farmers and the National Agricultural Advisory Service (N.A.A.S), to help the Ministry keep in contact with the current situation.

The N.A.A.S. was a new development, or perhaps one should say it was an upgrading of the old service which looked after and advised on the Veterinary and milk hygiene aspects of farming. It was based at eight Universities, and four Agricultural Colleges in England and Wales. The new advisory service took on experts to cover all aspects of food production and keep up with the latest research. This research was carried out on demonstration farms bought by the government in various districts, which meant that they could show the way forward in every geographical region of the country. Demonstration farms would be the best way to explain what was in the mind of the government.

Way back in 1923 the government was persuaded to extend the agricultural economic work then being done at Oxford University to all the Universities and Colleges doing other advisory work in addition to teaching. This new development had crept along during the depression but was re-invigorated during and after World War 2 and again at the beginning of the peace. Through farm costings the whole economic story of British farming was at last being put under the spotlight. Not only did it set out to help agriculturalists, but was a very useful tool to the nation which was paying out a great deal of money in the form of subsidies. By 1956 the food produced increased by leaps and bounds. The government used this arm of the advisory service to put the squeeze on farmers in the annual price reviews.

The farmers countered this development by improving the N.F.U. and it was this organisation with its economic expertise that sat down with the Government each year and argued out the prices for the coming year. Gradually an efficiency factor was brought in to the argument and this was making the farmer realise that he had to strive to make his farm better able to cope with what was in effect a price reduction each year. The 1960's heralded a change to bigger farms and much more mechanisation. Farm Labour was leaving at a great rate for a better life in the booming factories and rebuilding works. However farmers were still going for maximum production irrespective of the marketplace.

CHAPTER 17. THE TIME OF NEW IDEAS AND IMPROVEMENT OF OLD ONES

Thanks to the motor car industry and the money it generated farmers were about to have the chance of exploring the world for new ideas. Lord Nuffield was the man who endowed a trust with money that was used to send one or two farmers abroad each year for a year to work and look at what the countries visited had to offer. The men who went were very practical and were already running their own farms to a very high standard. Whilst they were away they were given money from the trust with which to employ staff to take their place at home.

On their return they were expected to travel around the land giving talks about what they had seen and learnt.

This was a great idea which benefitted the industry over the years especially in mechanisation and livestock management. Some of the first countries visited were New Zealand and the U.S.A. - the U.S.A. because it had begun to master the use of machines in farming well before the war, indeed way back when the prairies had first been tackled. Being far away from the markets and labour they had to fight hard to make money out of cereals and beef production.

New Zealand farmers on the other hand, were the worlds' experts in the management of grass for sheep, and milk production. They to, being far away from the market place, needed all the expertise they could muster. Their governments helped them with grants and other handouts, simply because farming was the nation's only money earner. So gradually back came ideas adapted by these practical farmers for British conditions. As the price squeeze hit British farming, those that were young and keen and had forgotten the sad days of the depression were able to overcome the price reductions. However many of the older chaps found it difficult to adapt or have confidence in the new era and they faltered and sold out to the 'go-getters'.

One man who took to heart the lessons of the Southern Hemisphere was a man called Rex Paterson. He took the bail-milking method, first started upon the downs from Hosier, the grass production methods from New Zealand and mixed all this with Ferguson's hydraulics and mechanised the ensiling of grass. It was the Victorians who discovered the advantages of conserving grass in this way. They were beaten by the hard work involved in handling wet grass. The chemical firm I.C.I. had tried to push the idea forward during the war and had proved how the feeding value of grass could be conserved by this method. The farmers would not take up the task again because of all the hard work involved. Paterson demonstrated what could be done. He used Nitrogen fertiliser in large amounts on his grass leys, producing good grass all the summer. He also produced high feeding value silage for the winter. This was stored in 'clamps' or heaps on the downs. The cows living upon the free-draining chalk downs fed their way through the silage heap, thus cutting out any feed handling by farm workers.

Paterson took over many acres of this down land, found which grasses were the most productive and began to show how cheaply milk could be produced by this simple system. He also adopted the New Zealand concept of not spending money on elaborate buildings. This all worked out well until dairymen decided that working in kinder conditions in warm factories was for them. Higher wages would be needed to keep them on the land so herd sizes would have to be increased. The additional mud and muck produced would not suit Paterson's simple system. British farmers had to change course.

They took the tools to hand and with generous government help began to expand and improve the livestock section of farming very quickly. Let us start with the Artificial Insemination of the dairy cow, started back in 1944 at Cambridge. The A.I. stations sprang up all over the country most of them run as a co-operative of farmers. The genetic make-up of the national herd quickly improved but one of the most dramatic changes was the sudden expansion of the Friesian breed.

The dominant breed at the time was the Dairy Shorthorn which had produced the milk for the war years. The Ayrshire breed was a close second. The Friesian was introduced from Holland to the Essex countryside around 1900 and began to increase in popularity due to its ability to produce milk consistently and a calf that was no slouch at growing into beef. This breed was excellent at converting grass into milk and produced a calf (usually cross-bred with a beef breed) which could be fattened on grass. The milk it produced was abundant but was low in butterfat. The Dairy Shorthorn started to be being crossed out to this newer breed and the Ayrshire which at one time looked like beating the rest to the winning post also faltered.

Many Friesians had been exported to the USA in the 1930s where a new breed was developed to become the Holstein, a cow that could be fed large quantities of concentrated feed such as maize and other grains.

An Ayrshire Cow

Dairy owners of the native breeds of Jersey and Guernsey benefitted from the higher prices paid for their rich creamy milk but it was a niche market.

A Jersey Cow

Those farmers who had had the wisdom to bring the black and white Friesian cow into East Anglia, some time after the First World War, began to grow rich on selling bulls to the new A.I. stations.

A Dutch Friesian

(Note modern breeds are de-horned to make handling easier)

At this time the cost of using the A.I. bull was cheaper than keeping your own bull and the very best in the land was yours for the asking. The beef breeds too came into their own via these stations and the testing of bulls for their rapid live weight gain began to be accepted by farmers.

It must be remembered at this time that the farmer's concept of judging cattle was only just weaning itself from the Victorian method of judging in the show ring for looks and fashion. The idea of looking for good milk yields and meat growth was fairly slow to become planted in their minds. Those who travelled the world and saw what was going on elsewhere began to get the message across but it must be said there was stiff opposition. It would be fair to say that from the late 1950's to the 1970's the rest of the world was improving its cattle at a faster rate than it had since the 1880's when many of the breeds had become established via the Herd Books and registration.

British breeders who over many years had been sending breeding stock all over the world were slow to catch up with this change. The world wanted new, more efficient cattle bred for different systems of rearing or climate. They wanted bigger framed beef animals that would cut out more meat per carcass. Two world wars and complacency had knocked the British breeder right out of the market place and he was not to get back his dominant place again until this century.
To compete he had in fact to go abroad himself and bring back beef breeds from the Continent which he improved before the still backward European farmers woke up.

Yet another dairy breed was to burst upon the scene from the mid 1960's and go like wild fire and take the place of the Dutch Friesian. It was big and could handle great amounts of concentrated foods and produce milk in very great quantities. It was the American Holstein and it had a great economic impact. As cattle produced more milk the herds became bigger and the number of dairy farmers shrank. The price of milk went down in real terms but the top notch dairymen prospered.

Artificial Insemination techniques did not stand still over this time. The discovery that you could freeze semen meant that it was at last possible to send it all over the world thus the very best genes spread more rapidly. When this happened the British A.I. stations began to find markets for this valuable material throughout the third world, and in the continent of Europe. Of course the fact that the milk recording which had only been sporadic before the war was now run by the Milk Marketing Board (M.M.B.) to cover the whole country meant its figures could be relied upon. The Holstein breed came to this country via the frozen straw route, initially from Canada where their small dairy men wanted small herds but very productive ones due to the introduction of milk quotas.

Cattle were not the only animals to be treated to A.I., the pig was being inseminated to spread the best over the country. During the 1950's pig herds had not followed the dairy herds in increasing in size. They had not moved very far from the 'cottager' era of management, a few sows and a boar around most farms, haphazardly managed to feed the public which up, until this time, was only looking for fresh pork as the bacon market had been collared by the Danes and their co-operative marketing methods, way back in the 1930's. An attempt was made in the early part of this period to counter this Danish attack, by the importation of the Landrace breed which the Danes had developed for the bacon market. This import came from Sweden for the Danes were reluctant to export their expertise to the country that could not get enough of their bacon. So the long lean bacon machine came to England, was snapped up, spread by A.I. but was unable to make much of a dent on the Danes due to the lack of marketing discipline.

By the mid 1960's pig herds had followed the trend to larger herds, kept in intensive purpose built buildings. To a large extent they were being concentrated in the main grain growing parts of the country such as Suffolk and Humberside. Pigs suffered as they always had done with the 'pig cycle'. That is because of their prolificacy they were able to increase numbers at a very rapid rate should the market demand more pig meat in the form of an increase in price. Unfortunately farmers are good at leaping on to a good thing when an excellent profit is showing and leaping off again when things take a turn for the worse. The timing of this cycle takes about three years and thus many farmers say of pig farming "Its muck or money". At one stage the pig men managed to have an arrangement with the government price fixing team that linked subsidy with the number of sows in the national herd. Thus the more sows the lower the guaranteed price and vice versa.

Gradually progress was being made in marketing and long term contracts to supply new bacon factories came into being. By the mid 1970's not so many pigs were sent to the weekly markets on speculation, to be bought and sold like stocks and shares, in the hope of making a good profit at the end of the day. As pig units increased in size and concentration in a given area so did the disease problems. These were often viral diseases which could devastate a pig farm in only a few weeks. In time the government was forced to bring in movement orders. Licences had to be obtained before any pigs could be shipped from unit to unit and this dramatically reduced the disease impact.

During the late 1960's and in the 1970's the future of the pig fell into the hands of the geneticists. It was they who produced faster growing breeds and greatly improved feed conversion rates, the amount of food needed to produce one pound of flesh. The speed of improvement and the greater expertise in handling genetic material meant that small groups of breeders pooled resources and started their own breeding companies.

Now we must take a look at how the poultry industry was adapting to the post war era. Before and for a few years after the war it would be unfair to call the production of eggs and fat chickens an industry. It was still more or less in the peasant era, with farmers having a few birds running around the farmyard and perhaps a few dozen cockerels fattened for the Christmas season. Turkeys too, by and large fell into this category. Most eggs were sent to the weekly markets to fetch what they might with a flush of eggs in the spring and nothing much to sell to a market by August. The hens of this period had not been bred to produce out of season as they do now. Thus the spring time saw a great increase in egg laying, falling off as the summer went on and more or less stopping in the winter months.

It was accepted that this was as Nature intended but did not do a lot to help the bank balance. One could preserve eggs when cheap in water glass chemical solutions but they could only be used in cooking.

Before the Second World War eggs were imported from as far afield as China and Russia but due to the political instability after the war these imports virtually stopped. Things were due for a change and change certainly came as the scholarship farmers began to visit the U.S.A. They learnt how to manage the humble hen so that eggs no longer were a luxury afforded only by the well off.

The spread of the National Electricity grid had a big part to play in this story. It meant that by 1955 most of the farms at last had electricity laid on. Thus they were able to put lights into buildings that housed the hens and extend the day, making the chickens think that spring was all year round. This, with improved nutrition, meant eggs began to come in increasing numbers. Profits were good and expansion took place at a very rapid rate. The Egg Marketing Board that began back in the 1930's was brought back to life to put some standards into the grading of the egg, and money into advertising the product. This was paid for via a levy on each dozen eggs put through a packing station. Their most effective advertisement was 'Go to work on an egg!'

Very soon another system became dominant in the production of eggs. Under this battery cage system hens could be kept far more densely. Food and water were kept constantly in front of the hens and the lights never went out. The system was designed so that eggs rolled away from the birds as they were laid and so were kept clean ready to pack without washing. Up went the numbers year on year at an immense rate and the price of the egg fell. By the 1970's it would be fair to say that unless a farmer kept some 30,000 birds in cages he was not going to make

much money. By the 1980's the number of cages needed to keep egg production viable rose yet again to around 50 to 100,000 birds all of which were fed and watered, cleaned out and eggs rolled away to be packed with out the help of anyone. This method eventually became an environment problem that the welfare of the hens was being debated by the public.
Like the pig, the old breeds of hen had been adapted by breeders to suit the new conditions and eggs per hen rose from 140 per bird to over 230. Not only did the birds lay more eggs, they needed very much less feed than the traditional breeds also helped by the fact that feed rations had also been improved by feed companies. Egg production became a very specialist industry and many farmers were being ousted by these specialist producers, often tied into the feed companies, who could command the capital to ever expand.

The so called cold war that came about, when Russia seemed about to expand its influence, had a part to play in the development of the poultry meat industry. At that time thousands of Americans were being billeted at air force bases all over the country to combat this Russian threat. They wanted their home comforts such as Southern Fried Chicken and these eating habits spread to the native population. The expertise together with the breed of bird needed to produce and meet the demand for this meat was imported from the States.

So came about the intensive rearing of the so called broiler chicken, in special houses. Again as in the egg production story the number of farmers catering for this market rose year by year until by the 1970's many had gone to the wall and the whole enterprise became dominated by big companies. It had its disease problems but in time they were overcome and the geneticists have ended up with breeds of chicken that take only some 40 days from hatching to slaughter and feed conversion rates that are getting very near to one for one.

At this stage in the mid 1960's farming in Britain was rather like a railway train. It had taken some years to get up speed. Despite letting go of trucks in the form of shedding labour to combat the government's price squeeze, the speed of the train never slackened. Farmers poured money into new methods. Thus output of everything increased and farmers were rushing to add to their acreage. Oddly the price of land had not moved very much over the years and good land could be bought at around £250 - £350 per acre. However, times were about to change and yet another era was about to establish itself on the farming world. It was the coming of the Common Market the European Economic Union.

Actually the Union was already in place brought about in the aftermath of the war. It was an effort by the old enemies to join the economies of their countries for the benefit of all and thus take away the rivalry of past decades, which had so often resulted in death and massive destruction. Britain failed to join when she first applied in 1960. That is a story in itself, which is worth following up, but not here. By 1970 the political climate had changed and we joined. We joined a club that was well on the way with ideas for the future both for the towns and countryside. It was going to be very difficult for us as a nation to change its way of thinking from leading an Empire to being part of an Empire. However the consensus was that we had to join for the sake of our industry which was failing due to the self rule of many old Empire lands.

Farmers, or most of them, relished the thought of joining. Many had made an investigation of European farming and realised that they were some 20 years behind our methods. It looked as if British farmers could out-sell their rivals on price any day. Certainly on the livestock front for were we not the New Zealand of Europe? Our Atlantic climate meant we could produce grass, the cheapest foodstuff in the world, and thus milk, beef and lamb.

Sadly the new era was not going to work like that. Protection was still to be the name of the game and would continue to be for some 25 years before there was any chance to take a market by climatic advantage or efficiency. When this state of play arrived in 1994 a new factor came along, the part farmers did or did not play in the environment of the countryside.

CHAPTER 18. 1957 THE TREATY OF ROME, THE WAY FORWARD FOR THE EUROPEAN FARMER

The Treaty of Rome is a very comprehensive document. It is the blue-print or plan that every country had to agree to carry out. It sets out the way forward in administration and political balance, payments into and out of a common pool. It is in my view rather like the Bible, open to interpretations that can and have been used for member countries own interests. However, it must not be forgotten that to bring all the countries involved into a common fold was not going to be in any way easy, it would have to be worked at rather like marriage, and would have to have a framework that would allow for change as the years went by.

In the early days back in the 1950's the fear of Communism was a factor in slowing the improvement of the farming structure. At the time Germany was encouraging their peasant farmers to work part-time on their small holdings so that they could also work in the booming metal industry. This was viewed to be the best way to keep them happy and not looking over the hedge at the East German experiment of collective farming. Much the same happened in Italy and France.

The Treaty of Rome gave a clear view of the way agriculture was to be developed by the creation of a Common Agricultural Policy (C.A.P). It was in essence, that it should be encouraged to improve its efficiency, so that Europe would be working towards self-sufficiency in temperate foods. The aim was that the income of farmers would over the years equate to that in other industries. In 1957 Europe was beginning to boom in economic terms and if peasant labour could be encouraged to work in the booming towns those farmers that remained on the land would also be able to reap greater rewards.

At the same time the old economic barriers between the club members would be removed and the market for everything would become a level playing field. With some 300 million people the market place was huge, as big as the U.S.A.

Here it would be as well to give an outline of the way the C.A.P. was intended to work. It is rather like a jigsaw puzzle, bits of the whole picture gradually put in as the years went by and bits replaced as old empire commitments faded.

Imagine the whole Community was surrounded by a wall (the tariff barrier) to food and other products. Within the wall a system was put in place that was intended to keep a constant price for the farmers produce. It was arranged by the Brussels commissioners who set up in each member country an intervention board whose job it was to buy in produce when the market was in surplus and put it back on the market when the produce was in short supply.

The intervention system began to work well during the 60's. The peasants from the French and Italian fields could not get off the land fast enough, to work shorter hours for far more money than their little small holdings could give. When the boom ended in the 1970's and the dole queues developed they were not so keen to leave. They fought by their vote to improve the lot of the farmer without the exodus.

From what has been written so far it does not take a lot of imagination to realise why most British farmers wanted to be part of the Common Market. It meant going into an organisation with a big rural vote, compared with being outnumbered within its own boundaries. A chance to help feed a very large population with food that, compared with the peasant farming of the Continent, could be produced cheaply. Bigger herds, bigger farms and more mechanisation must mean more

profits. In addition the British tax payer was intent on putting an end to the expensive subsidy system though they had no idea what was to come.
The British government could see that by joining the C.A.P. they could get off the hook of the ever burdening subsidy payments. They did not realise that the rural vote within Europe was very big compared with that in the U.K. The atmosphere within the other countries of the other voters was that of never to starve again, as they had done in the not too distant past.

Britain was the first to join the original six, followed by Ireland, Spain and Portugal. Joining was phased over a few years. The vision of the CAP was that joining countries could get monies out of the Community to refurbish their Agriculture so that at the end of the transition period the playing field would be similar. In Britain's case this meant that what was already the best producer within Europe improved its efficiency even more. The overall result was an increase in the amount of produce that came off the land within the EU. Up went grain production, milk and meat. The intervention stores were flooded with all this food leading to butter, beef and dried milk mountains. Then along came Spain's entry sucking in money to improve its farming and within a few years adding to the problem which eventually overwhelmed the Intervention system.

British farmers' dreams of pouring cheap food into the Continent therefore did not last because the EU politicians had to back pedal on their vision of supporting members until their farming had reached a common level. The governments of each member country were being forced to protect their own farmers irrespective of the overall EU plan, in order to keep the local vote. The cost of keeping the rural man happy rose to astronomical figures and eventually changes had to be made. It took some 20 years for this stage to be reached.

The howls of the French and German farmers in particular began to worry their Governments. As Brussels tried to curb intervention prices their farmers, being less efficient, suffered. This meant for a few years the Treaty of Rome was flaunted and money was put into these farmers' pockets, for votes, which again caused over production. What a mess, especially if you take into account the intricate system that was put into place to counter exchange rate variations, so that no farmer had an unfair advantage over others in other countries.

To try and overcome the overflowing intervention stores and curb the cost to the taxpayer Brussels began to export this produce, but to do so at world prices they had to subsidise them. The export of milk powder, grain and beef was not welcomed by other world farmers outside the E.E.C. The quantities grew so much that it distorted the true price of food on the world market and started to upset the balance of true trade. Many of the poor third world countries, who were always below the bread-line as their own production was so low, found the dumping of this excess food onto their markets far from helpful. In the long term it did exactly the opposite. It did not pay the third world chap to try and improve his farming when subsidised food from Europe cut his market to bits. It must be remembered that all this took place over some 20 years and could not go on, something had to give. G.A.T.T. was the wedge that began to dismantle this intricate system. In 1995 the first winds from G.A.T.T. began to blow but at first only slowly giving a chance for the world to adapt. Europe was not alone in sustaining its farming population.

In the year or two before the G.A.T.T. agreement came into force, Brussels had begun to put a measure into place that had a dramatic effect on the surpluses. This was at a cost, but not as great as the old system demanded. The set-aside arrangements (pioneered in the U.S.A.) came in. Not only did it restrict the amount of ground a farmer could plant to cereals, it also put a restriction on numbers of livestock that could be kept. Back in 1983 milk quotas had been introduced but even these had been set too high and had to be reduced in 1994.

The mountains of food began to melt like butter in the sun, but other factors that had been waiting in the wings concerning the environment and farming began to come to the fore, and are now only beginning to affect the farming world. This is the future.

CHAPTER 19. WHAT OF THE FUTURE FOR BRITISH FARMING?

This is a very difficult question to answer. In theory, assuming in the current jargon, world farmers are all playing on a level playing field then all should be well. British farmers will have to cooperate in many ways to become big players in the food industry. Producing food in penny packets will not do. That does not mean that some farmers will not find a local, niche market to supply, including organic produce. However it seems that this will only supply some 2% of the nation's needs and that would be for those who can pay for the extra cost of production. If you are rich you will not be concerned by the cost of food. However nobody wants to pay more than they have to for food and in the past we had been subsidised by the British Empire and the export of machinery.

Now the Empire has gone and the world is coming together. The odd thing is that the Europeans seem not to have followed this path. They still appear to look to their local markets for fresh foods. But is this changing? It is now 60 years since Europe was starving, since then the new generations have, in the words of Harold Macmillan the 1960s Prime Minister, 'Never had it so good'.

The EU was formed to help prevent wars in Europe, provide equal opportunities for all and for there never to be starvation again, in Europe at least. Until recently, it used subsidies to help ensure the stability of food production. This in turn created the inevitable surpluses which were then unloaded on world markets which then had a detrimental effect on those parts of the world with marginal farming.

However this is now changing. In future farmers will have to compete in the marketplace as other industries do, and not rely on those subsidies or at least this is what should happen. That is not to say that subsidies no longer exist but now tax payers money is being channelled in the direction of the environment of which much is tended by farmers.

Our supermarkets are always chasing in the world for cheap food produced by low-waged workers (at least in comparison to within the EU) to ensure they maintain the profits demanded by their shareholders. Eventually the world's labour will be paid more so if they can hang on for another few years then they are likely to be able to more than compete with the rest of the world, but this may be asking too much. Small British farmers, especially milk producers, are now going out of business. Unfortunately, unlike most industries, farming cannot change production methods overnight. The men doing the work cannot be trained in just a matter of weeks. When dealing with living things, not only livestock, but plants and seeds, several years of training and experience are needed.

As has been shown through the past centuries, once the farming world has been undone by a ruthless market place, it will be almost impossible to get it going again without a huge amount of effort and a lot of money!

What is needed is stable markets and controlled surpluses so that the world's soils can be tended properly. As far as Britain is concerned, as a very over crowded island life could become very difficult as it buys a vast amount of food from around the world. As the world gets hungrier, Britain will have to work harder to live. As other nations become richer so the price of all foodstuffs will rise. When this happens and if farming in Britain is not sustained, will Britain find it has a wasted asset? Time will tell.

Appendix 1 – Explanation of the 'loss of common rights' and the Enclosures Act

This is a very complicated subject. It started in the time of Henry VIII when there was a very valuable export market for wool. The big manor estates where demolishing whole villages to create sheep runs. This began to worry the government because land for growing cereals was decreasing with the possibility of famine. Thus land was enclosed to keep out the sheep by law.

Up until the 1700, as Landlords could see that the population was increasing and with new farming practices beginning to appear, they could see money being made. This meant that many began to enclose land with the agreement of their tenants and the reduction of the common land. Also the big open fields which had covered in cultivated strips began to go. As time went on more and more land was being enclosed and this was beginning to worry the government as unrest was beginning to appear among the commoners. To try and put the enclosure on a more legal base the Enclosure Acts came into being. These acts were fraught with problems and the fact that most of the Lords and House of Parliament was composed of land owners did not help.

In the 1840 many labourers, ex-commoners, revolted over low wages and the Swing Riots developed in the south of England.

Appendix 2 – Some Facts and Figures

Measurements

1 bushel is a volumetric measure equal to 8 gallons (about 36kg). Up until the 1890's bushels were always used to show output. This could be inaccurate as the weight of a bushel could vary from season to season, and area to area. Tons then took over, 1 ton being 2240lbs. These days metric tonnes are used - 1 tonne is 1000 kg

A bushel of wheat is about 50 kg, barley about 45kg and beans about 54kg

Hectares = 2.471 acres. An acre is 22 yards by 220 yards

Improvements in yields over time

Around the 12th century as far as we can tell the yields were as follows

Wheat 8-16 bushels or 400- 800 kg per acre
Barley 12-24 "
Beans 6-12 "

Data from the Ministry of Agriculture

	Milk litres per cow per year	Eggs per hen per year
1946	2291	120
1980	4716	248
1990	5895	310

Crops (tonnes per Hectare).

1900	Wheat 1.9	Barley 1.9	Potatoes 12.3
1946	Wheat 2.4	Barley 2.2	Potatoes 17.6
1980	Wheat 5.9	Barley 4.4	Potatoes 34.5
1990	Wheat 7.3	Barley 5.3	Potatoes 41.4

Of course it must be remembered that weather can have a good or adverse effect on yields.

A further indication of changes in the farming story:

Data from the Early Agricultural June 4th Returns

1867 – 8.7 million cattle.

1900 - 11.7 million cattle, the increase due to the loss of 3 million acres of wheat returning to grass.

Import data from the Ministry of Trade

1880 to 1885 500,000 cattle imported during from Russia and Canada

1907 Meat imports from:

Argentina 215,422 tons – Beef Products

Denmark 101,778 tons – Pig Products

Australia 51,700 tons of Beef and 35,000 tons of Rabbits

New Zealand 121,020 tons - Lamb

USA 305,415 tons – Mixed.

Census Data – UK Population, millions.

1881 10.5

1847 19.9

1891 33.4

1900 34.4

1930 45.8

1960 52.3

2012 63.8

Development of Farming Education

1790 Faculty of Agriculture started at Edinburgh University.

1834 First Highland show started in Scotland.

1839 Establishment of the Royal Society of England and also the first Bath and West Show.

1866 Chamber of Agriculture created to promote Agricultural interests in Parliament

1906 The Government gave £11.5 million to be spread among 20 colleges and institutions.